Rationality, Games, and Strategic Behaviour

Collected Papers in Theoretical Economics

Volume II

Rationality, Games, and Strategic Behaviour

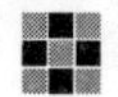

KAUSHIK BASU

OXFORD
UNIVERSITY PRESS

OXFORD
UNIVERSITY PRESS

YMCA Library Building, Jai Singh Road, New Delhi 110001

Oxford University Press is a department of the University of Oxford.
It furthers the University's objective of excellence in research, scholarship,
and education by publishing worldwide in

Oxford New York

Auckland Cape Town Dar es Salaam Hong Kong Karachi Kuala Lumpur
Madrid Melbourne Mexico City Nairobi New Delhi Shanghai Taipei Toronto

With offices in

Argentina Austria Brazil Chile Czech Republic France Greece
Guatemala Hungary Italy Japan South Korea Poland Portugal
Singapore Switzerland Thailand Turkey Ukraine Vietnam

Oxford is a registered trade mark of Oxford University Press
in the UK and in certain other countries

Published in India
By Oxford University Press, New Delhi

ISBN 019 566762 X

Typeset in Adobe Garamond 10.5/12 by Jojy Philip, New Delhi, 110 027
Printed by Pauls Press, New Delhi 110 020
Published by Manzar Khan, Oxford University Press
YMCA Library Building, Jai Singh Road, New Delhi 110 001

Contents

1 Introduction

1.1. By Way of Preface

Modern theoretical economics, as we know it today, originated in the late nineteenth century. I am referring here to the works of Leon Walras, Stanley Jevons, Vilfredo Pareto, and their contemporaries, though it would be wrong to ignore some of the precursors, such as Herman Heinrich Gossen, whose book is now considered a classic but had remained unknown till Jevons 'discovered' it almost a quarter of a century after its publication in 1854. These works gave rise to the 'marginalist-analysis' paradigm and general-equilibrium theory, which would dominate our discipline for the next hundred years and more.

This theoretical structure was built on one overarching simplifying assumption, which kept interaction between human beings manageably compartmentalized. Each person's welfare was affected by certain society-wide variables, such as prices, that prevailed in the market but which no single agent could alter. Prices rose and fell in response to large numbers of agents taking certain actions, with no single individual having any discernible impact. This meant that no single agent's behaviour could directly affect another agent's welfare. There were exceptions to this, of course, most notably in the work of Antoine Augustin Cournot published in 1838; and even some of the 'marginalists' themselves had considered deviations from this assumption of price-taking behaviour. But if there was a central paradigm that emerged from the profusion of economics writing in the late nineteenth century, it was that of general competitive equilibrium analysis, where this assumption of no direct interaction prevailed.

The marginalist and general-equilibrium school taught us a lot and has given us theorems and methods of thinking about the economy which will probably be with us for as long as recognizable human civilization lasts. But, as often happens, through force of habit, economists trained in this tradition began to resist any suggestion that not all parts of the economy at all times fit this description. So, though marginalist analysis is extremely important and provides tools essential to our craft, it—or rather the unwavering commitment of mainstream economics to it—became a hindrance. Unfortunately, those who tried to deviate from this main paradigm of the time did not have a consistent theoretical alternative to offer.

The rise of game theory in the late twentieth century has at last changed this. When John von Neumann and Oskar Morgenstern's seminal book came out in 1944, what may have looked like a mathematical curio, to be savoured by a few specialists but of little consequence to mainstream economics, turned out to be the basis of the second major revolution in theoretical economics. What has emerged is more than the apparatus of pure game theory; it is a different way to view interpersonal interaction in economic systems and is often described by the more general term 'strategic analysis.'

The 'strategic-analysis paradigm' recognizes that, while there may be some variables which no *one* agent can influence but which can affect our welfare, there are also ways—indeed several ways—in which individuals can directly affect one another's welfare. This immediately means that when one agent thinks how of to maximize his welfare, he has to keep in mind how this may affect another person and change her behaviour. Game theory provides a general analytical tool for theorizing about parts of the economy where these strategic considerations loom large.

This shift in paradigm has brought with it important changes in our normative thought (or, more accurately, the central tendencies of normative thought within the profession) and view of how policy should be crafted and what governments should do. One of the most important casualties has been the belief that individual rationality is sufficient to generate social optimality. This belief, which sprung up from a loose reading of some theorems of general-equilibrium theory and welfare economics, led to some important corollary 'folk wisdom', such as the need to keep government away from all market activity. The rise of strategic analysis showed up the fallacy of such 'market fundamentalism'. What is emerging now is a much richer and realistic understanding of markets, efficiency, and development. This is based on a rejection of both market fundamentalism and market bashing. It is now clear that the dilemma is not one of choosing between government and market. Both are needed in a modern society. For one, government has a major role to play in ensuring that markets run effectively. And well-designed markets, which recognize the importance of incentives and leave plenty of room for free contract, are important for the well-being of citizens and the economic progress of nations.

To reach this position has not been easy. It required a lot of hard analysis and painstaking research, some of which may have, initially, appeared abstract and unrelated to the real world. Indeed some of the abstract logical problems of strategic analysis are so interesting as pure philosophical puzzles that it is easy to lose sight of the ultimate problem on hand.

My own interest in strategic analysis has been driven by both these motivations. There were some problems that were aesthetically and philosophically so interesting that trying to solve them was, for me, an end in itself. Once I began working on such a problem, its implications for economic policy or human development would be, I must admit, far from my mind. In any case, I find the belief of some economists that every research should have a policy implication pedantic, to say the least. There were other problems that arose from the more down-to-earth concern of how best to craft policy or how best to understand some practical real-life economic problem. In this collection of my papers published over a long period, I have drawn on both these kinds of papers—the abstract and the quotidian. The common, binding matter of all the essays included here is the methodology. They all have something to do with strategic interaction between human beings. Some of them are explicitly problems of game theory. But some entail merely game-theoretic analysis. Indeed, game-theoretic analysis had interested me well before I learned game theory.

Some of these papers talk of possible outcomes in strategic models that may not be 'sub-game perfect' or even Nash. There are some economists who would consider such analysis wrong. I include these essays because I disagree with such a verdict. With every mainstream tendency arises its own orthodoxy. And there is in fact a new orthodoxy in economics that treats 'wrong' and 'game-theoretically wrong' as equivalent. I believe, on the other hand, that some of the understandings that we reach, using common sense, have the germs of new ideas and methods, even though these may not be articulated fully and formally. There are situations where a non-Nash outcome seems reasonable. To predict such an outcome is not a mistake (except in the eyes of the beholder who reflexively treats a 'predicted outcome' as another term for 'Nash equilibrium').

Some of my earliest interest in strategic analysis arose from my student days at the London School of Economics and, in particular, the lectures given by Amartya Sen. Hence the essays in this volume span a fairly long period of time and reflect some of my changing interests. Over this same time I have worked in tandem on development economics and those papers are published in a companion volume, *Collected Papers in Theoretical Economics, Volume 1: Development, Markets, and Institutions* (Oxford University Press).

1.2. Rationality and Social Norms

The concept of human rationality is fraught with intellectual pitfalls. For over two hundred years, social scientists and philosophers have tried to grapple with it and have made mistakes, ranging from proposing tautological definitions to logically inconsistent ones. At one level, game theory is all about rationality in interactive situations. The rise of game theory has helped us take huge strides towards a better understanding of rational behaviour. But puzzles and paradoxes remain. The two opening chapters of this volume deal explicitly with some of these basic puzzles. The chapters are very different though. The first one, 'The Traveller's Dilemma: Paradoxes of Rationality in Game Theory', is proper game theory, even though its aim is to challenge some routine thinking in game theory. The second chapter, 'On Why We Do Not Try To Walk Off Without

Paying After a Taxi Ride', written more than two decades ago, deals with a much more practical social problem concerning human rationality.

The first paper described a new game, that I called the 'Traveller's Dilemma'. I constructed the game deliberately to demonstrate the conflict between our intuition about rationality and what formal game theory has to say. In the Traveller's Dilemma all the formal definitions of game theory—Nash equilibrium, strict equilibrium, rationalizability—predict the same outcome in which the players fare poorly, but our intuition rejects the game-theoretic prediction and leans towards outcomes where the players do much better for themselves.

How does one reconcile the conflict between intuition and game theory? The answer depends on how we interpret the question and there are two significant ways of doing so. The first way is to treat this as a practical question: Why, in reality, would players play differently (and we know from laboratory experiments done with the Traveller's Dilemma that they do play differently) from the prediction of game theory? On this we have a plethora of answers. Human beings may not be maximizers of income; they have emotions of altruism, empathy, and generosity; they may treat small gains as no gain; and so on. On this there is a substantial literature and we have seen similar arguments being made in the context of other (extensive-form) paradoxical games, such as Reinhard Selten's Chain-store game or the Centipede. The harder question is the introspective one: Even if we *assume* that players maximize their personal incomes, an intuitive analysis of the Traveller's Dilemma suggests that they will violate the logic of game theory. How does one explain this? My paper gives no formal answer to this (merely sketching instead possible avenues of inquiry) and so its contribution must be viewed as that of helping us raise this troubling question in the context of *normal-form* games.

The second paper deals with a more worldly difficulty concerning rationality. It constructs an imaginary, though not unrealistic, example, which calls into question the widespread presumption among economists that individual rationality is all we have for bringing about social order and efficiency. What this presumption means is that, when we see bureaucrats taking bribes and not doing their jobs properly, this has to be thought of as either inevitable or a fault of the design of bureaucratic incentives and pay. Or, to put it in yet other words, the only way to make sure that, every time a judge is called upon to give a verdict, he will give the right verdict is to design a system of judicial pay and incentives such that he will find it in his *self-interest* to give the correct verdict. Individual integrity is entirely dispensable in this view.

The example in my paper is meant to challenge this view. It tries to persuade the reader that traits like personal integrity or conformity to social norms, which get so little mention in economics, are essential for understanding the order that we witness in society (to the extent that we do). The argument in the paper is not watertight. A skilled game theorist can, in the context of the example, construct sufficiently (and that too not very) complex arguments, which show how individual rationality is sufficient to create order and make people be honest, keep promises, and so on. My own contention is that, though there may be such an argument, what the example *suggests* is in fact true, that a modicum of personal integrity is essential for social order and the efficient functioning of society.

1.3. Games and Equilibrium Behaviour

Part II of the book is a collection of papers that deal with standard game theory, delving, in particular, into questions such as how to *model* rationality and what is the right solution concept to use in analysing a game. The last of this bunch of papers, entitled, 'Information and Strategy in the Iterated Prisoner's Dilemma', reproduced here as Chapter 7, was the first to be written. I wrote it when I was a PhD student at the London School of Economics. I had no training in game theory then and the paper emerged from a focused enquiry into a specific problem that we were introduced to in the course of a regular classroom lecture by Amartya Sen. This is the now-familiar problem of backward induction in the context of the prisoner's dilemma. The conflict between the logic of backward induction and intuition had troubled me and the paper emerged from my attempt to solve the 'paradox'. It led me to discover, for myself, the idea and significance of higher-order knowledge structures (statements like 'I know that you know that I am rational') and, in embryonic form, the concept of common knowledge; and in that paper I spelled these ideas out as clearly as I could in ordinary English. The paper was actually culled out from my PhD thesis, 'Revealed Preference of Government' (London School of Economics 1976).

I think I was lucky I began examining game-theoretic ideas before I learned any formal game theory; and this early encounter with a paradox, shaped some of my subsequent research. One such subsequent paper is 'On the Existence of a Rationality Definition for Extensive Games' (Chapter 4 in this book). This chapter refers to the same paradox that I grapple with in Chapter 7. In this chapter I show how the paradox can be transformed into an impossibility theorem. I believe that, in general, paradoxes reflect faults in our thinking. In other words, reality has no paradoxes. For instance, our tendency to believe that something exists (without making sure that such is the case) and then to search for that something's properties, or to subscribe implicitly to an assumption which is internally inconsistent and then to pursue its implications, are frequently the basis of paradoxes. So being able to come to terms with this particular one concerning solutions of games, by establishing a formal impossibility theorem, was particularly satisfying to me, providing confirmation of my general belief about paradoxes.

The idea behind the chapter is this. A solution concept (or, crudely speaking, the predicted set of possible outcomes of a game) embodies our understanding of rational behaviour in strategic contexts. But this idea is seldom formalized. In this paper I specified axioms which a solution concept ought to satisfy. Implicit in standard game-theoretic analysis is the assumption that no matter what players do in any period, all players are assumed to be rational from thereon and this is commonly known. In other words, in mainstream analysis, you can tear your hair, throw your money out of the window, and play so foolishly that everybody laughs but you will not be able to convince the other players of your irrationality. I introduced one axiom that disallowed this: If a certain set of actions is construed as rational, then any action outside of that set, taken by a player, must reveal him or her to every other player as irrational. The theorem that forms the heart of the chapter is the claim that this, along with the other seemingly reasonable axioms, creates an impasse. That is, there is no solution concept (not Nash, not rationalizability, no known or unknown concept) that satisfies these axioms.

This theorem is often misunderstood. It does not say that anything is possible in a game, for the claim 'anything is possible' is also a solution. It is the solution that consists of all available strategies. The theorem, on the other hand, says that there is no set of strategies that we, the analysts, can predict as the collection of possible outcomes. This fitted in well with my innate inclination towards scientific skepticism. I feel we too often reach too many conclusions; that we, human beings, are prone to err on the side of having too many beliefs, rather than too few.

To return to the theorem, note that it does not say that games cannot be solved but simply that they cannot be solved while holding on to the axioms I use. So, if one does not like the negative result, the lesson is to move carefully to see which of these axioms we may want to drop or whether we should think of the very idea of solution concepts differently from that in the standard literature.

It is this latter route that I took in my joint paper with Jorgen Weibull: 'Strategy Subsets Closed under Rational Behaviour' (Chapter 5 in this book). Here we followed a small literature and argued that solutions of games should not be thought of as collections of singletons—or a single strategy on the part of each player—but as an innately *set*-valued entity. This led us to develop a solution concept called *curb*—closed under rational behaviour—which has the attractive property that if a player believes that others will not play outside their specified sets, then he will have no reason to go outside his specified set. This solution concept has turned out to be of use to those working on evolutionary games and evolutionary dynamics.

The third chapter in this section, 'Group Rationality, Utilitarianism, and Escher's Waterfall' (Chapter 6 in this book), deals with a more specific question, rather than general ideas of solution concepts and rationality. The problem that is discussed here is of some philosophical interest and it is the philosophical problem that had intrigued me for quite some time and game theory came to the rescue much later. The problem is one that many moral philosophers have been concerned with.

To understand this, consider a society in which every human being is totally committed to utilitarianism (though it is fine to think instead of any agent-non-relative moral criterion). Hence, all agents have the same objective function. Therefore, in uttering any sentence an individual will not be concerned about its truth-value but simply whether it contributes to general happiness. Thus lies, which enhanced aggregate human happiness, would be common in such a society. But if everybody behaved like this, surely language would cease to have some of its efficacy, because we would not believe in one another. Some philosophers had wondered if this damage to language could be so great (and, through that, damage to the fabric of our lives made possible by language so great) that human beings would end up being worse off in terms of the utilitarian criterion. In other words, could each person's pursuit of utilitarianism in every instance of decision making make humankind worse off in utilitarian terms itself?

To take another example, suppose that runners in the Olympics are committed utilitarians who slow down and let others overtake them on the tracks whenever they feel that that will generate more aggregate utility than their winning the race. It does seem that such behaviour, even if on each occasion it enhances aggregate welfare, will eventually mean that no one will go to watch the Olympics (watching a group of moralists making way for one another can hardly be an exciting spectator sport) and this damage to the joy of Olympics may be great enough that in the end society will be worse off in terms of the same moral criterion that it were out to maximize, namely utilitarianism. The game that I called 'Escher's Waterfall' shows that this is indeed possible. Of course it is an unusual game and so in some sense it illustrates the kind of situation where this self-destructive pursuit of morality is possible.

This game has important implications for moral philosophy. It pushes us towards deontological ethics and, more minimally, gives us reason to abandon act consequentialism. It says that there are certain things in life that we should do, without concern for moral consequence (at least so long as the moral consequences are not too large), that when you run in a race you should try to win it, when you speak you should try to convey information, when you play chess you should try to beat your opponent. There may be special cases for deviating from these rules but to do so systematically to achieve certain moral consequences would be self-defeating. As a result of this paper I have now come to believe that there are domains of life where one's moral decisions should be guided by deontological principles, though with an eye on consequences lest they be too large to be ignored.

I have found the model in this paper useful for other kinds of research, such as evaluating the role of contracts in labour markets. It also relates closely to the kind of philosophical venture attempted by Derek Parfit, where he tries to show that there may be certain kinds of actions where the moral worth of a class of actions may not be equal to the sum of the moral worth of each action in that class. Escher's Waterfall illustrates the sense in which this is true and has policy implications for legislating against workplace exposure to health hazards, even when workers agree voluntarily to expose themselves to high risk (maybe because they consider the financial compensation adequate) and other matters of labour standards and rights.

1.4. Industrial Organization and Strategic Behaviour

Of all the fields where game theory has found use, industrial organization is the most prominent. It was the rise of game theory that breathed new life into the field of industrial economics. This has not been an unmixed blessing. With game theory has come a multiplicity of predictions for strategic industrial situations, such as oligopolies. Many lament the disappearance of old certainties, where, essentially, the Cournot equilibrium was the only serious candidate when it came to predictions. But the fault, surely, does not lie with game theory, but with the world. Game theory is a method that alerts us to the risk of assuming certainty in a world that does not permit this. The problem of multiple possibilities is not a construct of game theory and strategic analysis but a statement of the way things are. And viewed in this way, game theory has helped rather than hurt us. We may now try, once again, to zero in on a narrower range of predictions starting from this richer method of analysis; and if we succeed we will have reason for greater confidence in our predictions.

I got interested in industrial organization theory rather late in my career and mostly as a spin-off of other interests. My paper 'Monopoly, Quality Uncertainty, and "Status Goods" ', reproduced in this book as Chapter 8, is a good example of this. There are many examples in life where one sees queues and waiting lists, where they seem unnecessary. There was a time when to buy a Jaguar car or a Bajaj scooter one had to book and wait for several weeks before one got delivery of the product. There are restaurants where one has to wait a long time before one is seated. Mainstream economics would suggest that the supplier of the product should in each of these cases raise price till the excess supply vanishes. In my chapter I try to show that if one person's demand depends on how many others are interested in buying the product then one could get such a persistence of excess demand in equilibrium, despite producers being rational. 'Status goods', goods where information is important but scarce and so one has to observe other people's behaviour to deduce information, are candidates for this phenomenon. Modelling this is useful not only in itself but also, as I tried to show in some later works, in understanding some other important real-world phenomena, such as the persistence of excess supply of credit that some nations face.

The last paper in this section, 'Why are so Many Goods Priced to End in Nine? And why this practice Hurts the Producers' (Chapter 14 here) also emerged out of everyday observation. Casual observation suggests that a disproportionate number of products and services sell for prices that end with the digit '9'. It is not uncommon to find shoes for Rs 899, vacation packages for £599, and burgers for ¢99. One easy answer that the marketing literature has suggested is simply that people ignore the last (and, therefore, less important) digits of prices. Firms take advantage of this laziness of the consumers and charge the maximum they can without observation. Hence, the nines. The explanation I provided was a little different. In my model also consumers do not always look at the last digits, but, instead of ignoring those digits, they assume those digits to be the average of what they actually are. In other words they indulge in a kind of 'rational' filling in of the blanks. This simple assumption also means that prices will end in nines but changes the welfare implications quite sharply. Now the producers are the losers, since no single producer can charge a non-nine price without losing money. In other words their choices get limited and they do worse in equilibrium.

The cluster of four papers, Chapters 9–12, belongs to more traditional industrial-organization theory. The paper 'Why Monopolists Prefer to Make their Goods Less Durable' (Chapter 9) constructs an argument why monopolists when they face a choice between making a good more durable and higher priced will have a tendency to stop short at an inefficient point where price and durability are less than optimal. I do not know if this is empirically borne out, but provide theoretical reasons for expecting this to be so.

My paper, written jointly with Nirvikar Singh, 'Entry-Deterrence in Stackelberg Perfect Equilibria' (Chapter 10), takes forward an agenda of research that was begun by Michael Spence and Avinash Dixit and studies strategies that incumbent firms follow to deter entry. Unlike in the previous literature, we assume the post-entry game to be a Stackelberg one in which the incumbent moves first and the entrant after that. This case is, in some ways, the most natural but was left unanalysed in the literature, maybe because of a presumption that it is uninteresting. We showed that that was not the case. Properly modelled, this had its share of unexpected results, with important implications for anti-trust policy.

Chapter 11, 'Collusion in Finitely Repeatedly Oligopolies', assumes that the standard oligopoly game is played in real time and is, therefore, best modelled as an extensive-form game. Given this assumption, a repeated play of oligopoly means a repeated play of a game that is itself an extensive-form game. Using this characterization allows us to get the interesting result of how even a finitely repeated oligopoly could lead to collusive behaviour, once again with policy implications for anti-trust legislation.

Chapter 12, 'Stackelberg Equilibrium in Oligopoly: An Explanation based on Managerial Incentives,' belongs to the literature on strategic delegation in the theory of the firm. This is a simple paper that endogenizes the decision of each firm in a duopoly whether or not to have a manager to whom all output decisions are to be delegated and whose pay structure is carefully designed to enhance the firm's profit. The paper attracted attention, maybe because it was also very convenient for addressing certain questions in agrarian-organization theory.

'The Strategic Role of International Credit as an Instrument of Trade', written with Ashwini Deshpande, and reprinted here as Chapter 13, is somewhat different, because it is situated halfway between industrial-organization and international-finance theories. My interest in this field arose from development economics and my earlier work on international debt crises. The international bank that lends money to a poor nation and multinationals that sell goods to the same nation are by no means price-taking small agents, at least not in the arena of interaction with a poor developing nation. Once these agents are modelled as strategic players in a game, some interesting and counter-intuitive results begin to emerge. The kind of policy that is analysed in this chapter is one in which more international credit is made available to a credit-starved nation. We show that this can actually lead to a lowering of welfare among the consumers of the credit-starved economy, the ultimate gainers being the multinational companies. It is not suggested that this will always or even normally happen, but this possibility highlights a danger that must be guarded against by international organizations designing international credit schemes with the idea of helping poor nations. This kind of strategic international finance has not received enough attention. I think there is a lot of real-world potential in this and am currently engaged in some related research projects.

1.5. Governments, Games, and the Law

Much of traditional economics treated governments as lying beyond the ambit of standard analysis. Government was viewed, in such models, as an exogenous agent. Once we break away from this assumption straightjacket, the life of the analyst, of course, gets more complicated but a huge and exciting research arena opens up. Not surprisingly, there is now a growing field of enquiry, by economists and political scientists, that studies the implications of 'endogenous government', that is, broadly speaking, a view of government in which the agents involved—the police, the ministers, the bureaucrats—are like other players, with their own aims and objectives and strategic considerations. Chapters 15–17 may be viewed as contributions to this research agenda. For me, these chapters turned out to be a first taste of a field of research that has occupied a lot of my time in recent years and culminated in my book, *Prelude to Political Economy: A Study of the Social and Political Foundations of Economics* (Oxford 2000). This book also sounds some notes of caution about over-zealousness in this inquiry.

In 'Notes on Bribery and the Control of Corruption' (Chapter 15), which I wrote with Sudipto Bhattarcharya and Ajit Mishra, we study the problem of controlling corruption where the agent entrusted with this task is himself or herself liable to corruption. We allow for the possibility that a more senior administrator appointed to monitor the junior officer is also liable to bribes and so is the next level and the next .... The chapter develops a method for solving this hierarchy of decision problems simultaneously.

The paper, 'A Model of Monopoly with Strategic Government Intervention', written jointly with T.C.A. Anant and Badal Mukherji, investigates the problem of taxing an industry where the taxing agent, namely the government, is also a player. The Nash equilibrium of this game turns out to be easy but with implications of some interest. It is shown, for instance, that a firm in this environment, faced with different technologies, will not always have the incentive to choose the most cost-efficient technology, thereby upsetting one of the cornerstones of standard industrial economics.

Chapter 17, 'On Misunderstanding Government: An Analysis of the Art of Policy Advice', is technically an easy paper but deals with a problem that is conceptually hard and quite central to appreciating economics as a discipline embedded in society and politics. This is basically on the link between the worlds of speech and action. Note that a policy adviser usually utters some words to the seeker of advice, who then has to take some action. Hence, central to understanding the role of an adviser (which is often the role that the economist himself or herself plays) is understanding the interaction between a speaker and an actor, where both may have their own motivations and interests. There is a now a small body of game-theoretic literature on this and this chapter takes this line of enquiry to specifically the question of how we model governments and their advisers.

The closing chapter of this collection, 'The Economics and Law of Rent Control', written with Patrick Emerson, is a more straightforward exercise in law and economics, using the method of strategic analysis. It examines a problem of enormous contemporary, practical importance, namely, the subject of rent control. This topic has spawned a large literature that is essentially empirical, with most of the theoretical assertions taking the form of folk assertions. Our chapter models this carefully, using standards methods of industrial market analysis, clarifying why rents are best left as a matter of voluntary contract, but also investigating the scope, admittedly limited, for legal and other government interventions in the housing market.

Part I
Rationality and Social Norms

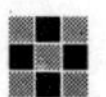

2 The Traveller's Dilemma
Paradoxes of Rationality in Game Theory

This chapter presents a parable which highlights the conflict between intuition and game-theoretic reasoning. One of the basic ingredients of analysis in game theory is 'backward induction'; but backward induction is also the source of some deep paradoxes (see for example, Binmore 1987; Pettit and Sugden 1989). Well-known games such as the finitely repeated 'prisoner's dilemma', Selten's (1978) 'chain store', Rosenthal's (1981) 'centipede' and Reny's (1993) 'take-it-or-leave-it' highlight this conflict between backward-induction reasoning and other kinds of reasoning.

Much effort has gone into trying to solve the problem. Virtually all of these efforts exploit the extensive-form structure of the above games or the fact that they are played over time. Thus Binmore and Brandenberger (1990) observe that these paradoxes arise because players in the above games can 'throw surprises' on one another by deviating from the path suggested by backward induction. If all moves by all players were occuring at a point of time, 'throwing surprises' would be inconsequential because it would influence no one's behaviour. Reny (1993) also locates the paradox in the sequential character of these games, arguing that the problem arises because 'during the course' of some plays, Bayesian rationality cannot be common knowledge.

This chapter demonstrates that the above problem is deeper, because it can arise in a single-shot game. This is done by constructing a paradoxical game—the 'traveller's dilemma'. The backward induction in the traveller's dilemma occurs at an introspective level. The standard suggestions for battling the

From *American Economic Review*, May 1994.

I am grateful to Cristina Bicchieri, Peter Bohm, Eric Van Damme, Martin Dufwenberg, Yvan Lengwiler, Ajit Mishra, Bob Rosenthal, Larry Samuelson, and Jorgen Weibull for useful conversations.

backward-induction paradox in, for instance, the repeated prisoner's dilemma (for example Bicchieri 1989) do not seem to be possible here. Hence this problem cannot be solved by attributing unusual knowledge structures at unreached nodes.

All intuition seems to militate against all formal reasoning in the traveller's dilemma. Hence the traveller's dilemma seems to be one of the purest embodiments of the paradox of rationality in game theory because it eschews all unnecessary features, like play over time or the non-strictness of the equilibrium.

2.1. The Parable

Two travellers returning home from a remote island, where they bought identical antiques (or, rather, what the local tribal chief, while choking on suppressed laughter, described as 'antiques'), discover that the airline has managed to smash these, as airlines generally do. The airline manager who is described by his juniors as a 'corporate whiz', by which they mean a 'man of low cunning', assures the passengers of adequate compensation. But since he does not know the cost of the antique, he offers the following scheme.

Each of the two travellers has to write down on a piece of paper the cost of the antique. This can be any value between 2 units of money and 100 units. Denote the number chosen by traveller i by n_i. If both write the same number, that is $n_1=n_2$, then it is reasonable to assume that they are telling the truth (so argues the manager) and so each of these travellers will be paid n_1 (or n_2) units of money.

If traveller i writes a larger number than the other (that is, $n_i > n_j$), then it is reasonable to assume (so it seems to the manager) that j is being honest and i is lying. In that case the manager will treat the lower number, that is n_j, as the real cost and will pay traveller i the sum of $n_j - 2$ and pay j the sum of $n_j + 2$. Traveller i is paid 2 units less as penalty for lying and j is paid 2 units more as reward for being so honest in relation to the other traveller.

Given that each traveller or player wants to maximize his pay-off (or compensation) what outcome should one expect to see in the above game? In other words, which pair of strategies, (n_1, n_2), will be chosen by the players?[1]

In order to answer this question it is useful first to express this game as a pay-off matrix. Observe that the above game could be thought of as having at least two versions, depending on whether the players can choose any real number or can choose only an integer. For most of the time I shall assume the latter, since that is where the main problem arises. When I assume the former, I shall refer to the game as the 'continuum version of the traveller's dilemma'.

2.2. The Paradox

At first sight, both players feel pleased that they can get 100 units each. To get this, each player simply has to write 100. But each player soon realizes that if the other player adheres to this plan then he can get 101 units of money by writing 99. But, of course, both players will do this, which means, that each player will in fact get 99 units. But if both were planning to write 99, then each player will reason that he can do better by writing 98; and so on. The logic is inexorable, and there is no stopping until they get to the strategy pair (2,2), that is, each player writes 2. Hence, they will end up getting 2 units of money each. This illustrates how backward induction, at the level of introspection, works even in a one-shot game.

It is easy to check that all standard solution concepts predict outcome (2,2). This is the unique strict equilibrium of the game, the only Nash equilibrium, and, in fact, the only rationalizable equilibrium. Yet it seems very unlikely that any two individuals, no matter how rational they are and how certain they are about each other's rationality, each other's knowledge of each other's rationality, and so on, will play (2,2). It is likely that each will play a large number in the belief that so will the other, and thereby they will both get large pay-offs. At one level the traveller's dilemma shares some similarities with the Bertrand duopoly, especially one in which firms choose prices from a grid; for instance, the set of integers starting from an integer above the marginal cost and up to the monopoly price. The best response structure of such a duopoly is similar to the best-response structure of the traveller's dilemma. However, that is where the analogy ends. In the Bertrand duopoly, if one firm chooses a price even slightly above the other's price, it earns zero profit. The penalty is nowhere nearly as severe for choosing a higher number in the traveller's dilemma. This is exactly what makes it plausible that players will choose large numbers in the traveller's dilemma. It may be possible, however, to construct a model of differentiated-products duopoly which is exactly analogous to the traveller's dilemma.

In the finitely repeated prisoner's dilemma, it has been shown that co-operation in the early games is possible if one uses the (single-shot) rationalizability criterion. In this game (2,2) is the unique rationalizable outcome. Observe also that, unlike in this game, in the centipede or the take-it-or-leave-it game the 'unwanted' equilibrium is not strict.[2] Hence in terms of formal analysis there seems to be no escape from (2,2).

But even knowing all this, there is something very rational about rejecting (2,2) and expecting your opponent to do the same. This is the essence of the traveller's *dilemma*. This is also the reason why escape routes which are made possible by allowing for irrationality or the expectation of irrationality (see for example, Kreps et al. 1982) are not of relevance here even though they may be important empirically.[3] It is not an empirical point that is being made here. The aim is to explain why, despite rationality being common knowledge, players would reject (2, 2), as intuitively seems to be the case.

2.3. The Possibilities

While I am unable to resolve the paradox, what follows are some possible lines of attack. Possibility 1 suggests a rigorous resolution of the problem for a special case, to wit, the continuum version; possibilities 2 and 3 should be treated as speculative rather than formal.

Possibility 1. The continuum version of the traveller's dilemma has an interesting way out by using an adaptation of the concept of curb sets, developed in Basu and Weibull (1991)—curb being an acronym for 'closed under rational behaviour'.

In the continuum version, each player *i*'s set of strategies is given by $S_i = [2, 100]$. Let T_i be a subset of S_i, $i = 1,2$. The pair (T_1, T_2) is defined as *curb* (actually 'tight curb' in Basu and Weibull 1991) if T_i is the set of all best responses of player *i* to *j*'s strategies in T_j, $i = 1,2$. In other words, the strategy s_1 belongs to T_1 if and only if there exists a strategy s_2 in T_2 such that player 1 cannot do better by unilaterally deviating from (s_1, s_2).[4]

A direct application of curb to the continuum version of the traveller's dilemma is not possible because there are no best responses in this game. However, here is a modified version of curb—I shall call it M-curb—which uses the idea of curb. Let (T_1, T_2) be called **-curb* if T_1 and T_2 are non-empty and, for all s_2 in T_2 and all s_1, in S_1, there exists r_1 in T_1 such that player 1 does at least as well by responding to s_2 with r_1 instead of s_1, and likewise with players 1 and 2 interchanged.

(T_1, T_2) is *M-curb* if it is *-curb and *individually* minimal, that is, there does not exist M_1 which is a proper subset of T_1 or M_2 which is a proper subset of T_2 such that $(M_1, T_2)$is *-curb or$(T_1, M_2)$is *-curb.

It is easy to see that if (T_1, T_2) is M-curb then $\max[T_j]$ is either 2 or does not exist, for $j = 1$ or 2. Here is an example of an M-curb set: [(90,100), (90,100)]. Hence, if each player commits to play in the open interval (90,100), then no player has the incentive to deviate. While this is a resolution of the continuum version, this cannot be taken as a resolution of the paradox, because the heart of the paradox does not lie in the technical matter of whether players are allowed to use all real numbers or not; I will now turn to the integers version.

Possibility 2. Though the traveller's, dilemma is a normal-form game, it nevertheless can be thought of as having the 'unreached-node problem'. To see this, begin by (a) defining rational play in the usual way and then (b) assume that rationality is common knowledge.

Since (2,2) is the only rationalizable outcome, it follows that that is what one should expect, since rationalizability is the consequence of (a) and (b). Now suppose player 1 wants to decide how he would do if he rejected playing 2 and went instead for a larger number. It is not clear that this question is at all answerable. If it is true that (a) and (b) imply that player 1 will choose strategy 2, then a world where (a) and (b) are true and the player chooses some other strategy may not be conceivable, and so such introspective experiments may not be possible.

One possible line of attack that this suggests is to argue that the implicit assumptions (a) and (b), which underlie so much of game theory, may by themselves be inconsistent. In Basu (1990) I showed that, in the context of games like centipede, the problem stemmed from assuming that rationality is common knowledge and that every game must have a solution. The method was to write down some properties of a solution, given that rationality is common knowledge, and to demonstrate that these properties cannot be together satisfied. However, there I made critical use of the extensive structure of the game. The traveller's dilemma challenges one to construct similar theorems without recourse to the sequence of play.

POSSIBILITY 3. To end on an optimistic note, I shall now consider a more novel line of attack. Observe that in the traveller's dilemma there cannot exist a well-defined set of strategies, T_i, excepting the special case $T_i = \{2\}$, such that:

(i) a rational player may play any strategy in T_i and will never play anything outside it, and

(ii) such a T_i can be *deduced* from an examination of the game.

To see this, suppose that T_1 and T_2 are such sets. Since player 2 is perfectly rational, he can deduce what the game-theorist can deduce. Hence, by (ii), he can deduce T_1. Let t be the largest number in T_1. Since player 1 will never play any number above t, it never pays for player 2 to play t. Hence T_1 and T_2 are not identical. But since this game is symmetric and T_1 and T_2 are deduced purely from examining the game, T_1 must be the same as T_2. This contradiction establishes that no such (T_1, T_2) exists.

Note that this whole exercise was for well-defined (that is, the usual kind of) sets. Hence, there *may* exist ill-defined sets that would work. There seems to be some *a priori* ground for believing that there may be an escape route here. Harking back to an idea that was touched on earlier, suppose that player 1 believes that player 2 will play a large number. Then, if player 1 were simply deciding whether he himself should play a large number or not, it would be in his interest to play a large number. Thus (large, large) seems to be a kind of Nash equilibrium *in ill-defined categories*. The ill-definedness is important here because if the set of large numbers was a well-defined set, one knows from the above paragraph that this argument would break down.

I am here interpreting 'large' in the sense of everyday language, which is different from the fuzzy-set-theoretic interpretation. The latter implies that the set of integers that are certainly not large is a well-defined or crisp set. The everyday use of the word 'large' clearly does not conform to this. Once this is taken seriously, many objections concerning the idea of Nash equilibrium in ill-defined categories, which immediately come to mind, cease to be valid.

Consider a question like this: 'if the other player is playing a large number, should I ever play a number 1 less than a large number?' Once one starts answering questions like this, the argument as to why (large, large) is a kind of Nash equilibrium will quickly break down. What I am arguing, however, is that the question like the one above is not permitted in this framework. Given the everyday use of the word 'large', 'a large number minus 1' is a meaningless term. All I am claiming here is that if a player is told that the other player will choose a large number and then asked whether he will choose a large number or not, he will say yes.

The use of imprecise categories does not mean forgoing rationality. What was argued in this subsection was that one way of holding on to the rationality assumption in the face of paradoxical games such as the traveller's dilemma may be to allow players to use ill-defined categories in doing their reasoning about how to choose in game-theoretic situations.[5]

NOTES

1. This game is a generalization of the prisoner's dilemma, since, if the travellers had to confine their choice to 2 or 3, we would have exactly the prisoner's dilemma. (A different generalization of the prisoner's dilemma occurs in Basu 1994.)
2. For different perspectives on the standard backward-induction paradox, see, for instance, Schick (1983), Sen (1985), Taylor (1987), Bonanno (1991), Borgers and Samuelson (1992), Dufwenberg and Linden (1993), and Hollis and Sugden (1993).
3. Confronted by a similar problem involving the iterated deletion of dominated strategies, Glazer and Rosenthal (1992) argue that players play co-operatively because they do not mind forgoing the small gains of non-cooperative play. This may be so in reality, but my problem stems from the belief that even players who are scrupulous maximizers would play large numbers in a game like the traveller's dilemma.
4. This is actually an imprecise definition of curb but it captures its essential idea.
5. For an ingenious related analysis, see Bacharach (1991).

REFERENCES

Bacharach, Michael. 1991. 'Games with Concept Sensitive Strategy Spaces'. University of Oxford. Mimeo.

Basu, Kaushik. 1990. 'On the Non–Existence of a Rationality Definition for Extensive Games'. *International Journal of Game Theory* 19 (1): 33–44.

——. 1994. 'Group Rationality, Utilitarianism and Escher's Waterfall'. *Games and Economic Behavior* 7: 1–9.

Basu, Kaushik and Jorgen Weibull. 1991. 'Strategy Subsets Closed Under Rational Behaviour'. *Economics Letters* 36 (2, June): 141–6.

Bicchieri, Cristina. 1989. 'Self-Refuting Theories of Strategic Interaction: A Paradox of Common Knowledge'. In Wolfgang Balzer and Bert Hamminga, eds. *Philosophy of Economics.* London: Kluwer.

Binmore, Ken. 1987. 'Modeling Rational Players, Part I'. *Economics and Philosophy* 3 (2, October): 179–214.

Binmore, Ken and Adam Brandenberger. 1990. 'Common Knowledge and Game Theory'. In K. Binmore, *Essays on the Foundations of Game Theory.* Oxford: Blackwell.

Bonanno, Giacomo. 1991. 'The Logic of Rational Play in Games of Perfect Information'. *Economics and Philosophy* 7 (1, April): 37–65.

Borgers, Tilman and Larry Samuelson. 1992. ' "Cautious" Utility Maximization and Iterated Weak Dominance'. *International Journal of Game Theory* 21 (1): 13–27.

Dufwenberg, Martin and Johann Linden. 1993. 'Inconsistencies in Extensive Games'. Department of Economics, Uppsala University, Working Paper.

Glazer, Jacob and Robert Rosenthal. 1992. 'A Note on Abreu-Matsushima Mechanisms'. *Econometrica*, 60 (6, November): 1435–8.

Hollis, Martin and Robert Sugden. 1993. 'Rationality in Action'. *Mind* 102 (405, January): 1–35.

Kreps, David, Paul Milgrom, John Roberts, and Robert Wilson. 1982. 'Rational Cooperation in the Finitely-Repeated Prisoner's Dilemma'. *Journal of Economic Theory* 27 (2, August): 245–52.

Pettit, Philip and Robert Sugden. 1989. 'The Backward Induction Paradox'. *Journal of Philosophy* 86 (4, April): 169–82.

Reny, Phil. 1993. 'Common Belief and the Theory of Games with Perfect Information'. *Journal of Economic Theory* 59 (2, April): 257–74.

Rosenthal, Robert. 1981. 'Games of Perfect Information. Predatory Pricing and the Chain Store Paradox'. *Journal of Economic Theory* 25 (1, August): 92–100.

Schick, Frederic. 1983. *Having Reasons*. Princeton, NJ: Princeton University Press.

Selten, Reinhart. 1978. 'The Chain Store Paradox'. *Theory and Decision* 9 (2, April): 127–59.

Sen, Amartya. 1985. 'Goals, Commitment and Identity'. *Journal of Law, Economics, and Organization* 1 (2, Fall): 341–55.

Taylor, Michael. 1987. *The Possibility of Cooperation*. Cambridge: Cambridge University Press.

3 On Why We Do Not Try to Walk off without Paying after a Taxi-Ride

3.1

This almost facetious question throws light on two conflicting claims about the order that prevails in society. The paradoxical claim (P) asserts that the pursuit of selfish aims by individuals is sufficient to generate social order. The unparadoxical claim (U) asserts that human adherence to certain commonly accepted values is necessary. Unless we define selfishness so loosely that it is difficult to conceive of unselfishness, (P) and (U) are conflicting claims. This is worrying because both occur within the social sciences: the former in economics and the latter in sociology. The purpose of this chapter is to demonstrate with an example, which is representative of a class of real-life situations, that it is the unparadoxical claim which is valid.

When we travel by taxi, we do not usually make an effort to run away without paying. Some economists have tried to model crime and marriage in terms of individual rationality, so they would no doubt explain our scrupulousness in terms of the probability of being caught and the agony of being jailed. In order to permit a relatively rigorous discussion let us place the problem at a greater level of abstraction.

An individual gets off from a taxi at a place where there is no one sufficiently near to bear witness as to whether he pays the fare or not; and in the absence of a witness, it is pointless contacting the police (with some police forces this is unconditionally so). Further, this is a large city and the passenger does not expect to require the services of this cabman in the future. Would the passenger try to walk off without paying? I think there will be no disagreement that, even in this situation, a vast majority of human beings would not choose to default.

From *Economic and Political Weekly* 18 (November) 1983: 2011–12.
I am grateful to André Béteille, Mrinal Datta Chaudhuri, and Amartya Sen for discussions.

How do we explain this, excepting in terms of our sense of values, or our morality, or custom (essentially something *beyond* selfishness)?

The economist's 'trained incapacity' does not allow him to give in so easily. I posed this question to a number of economists, with one or two exceptions, the response was more or less the same: while the passenger's sense of values may indeed be the cause of his adherence to the law, the act could also be explained purely in terms of rationality. Taxi drivers are often quite strong (which, judging by my small sample, is by no means the most vulnerable assumption). So if the passenger tried to button up his pockets, the taxi driver would in all likelihood either himself assail the passenger or gather people and try to ensure payment. It is this risk which makes the passenger 'behave'. This is a plausible argument and let us accept it.

But the moment we do so we land ourselves in a problem. The crack, however, appears elsewhere. It is now the taxi driver's rationality which becomes questionable. Why do we expect him to retaliate against a passenger who tries defrauding him, and to attempt recovering the fare at the risk of, or despite, the unpleasantness of a scuffle? To me, the most plausible reason seems to be his injured sense of fair play or anger at his customer's violation of social norms (no doubt catalysed by the fact that he is at the receiving end). But to admit this is to grant the role of commonly accepted values—no matter how indirectly—in the prevention of anarchy. This leaves only one way out: to explain the taxi driver's response in terms of his selfishness. To do so, one would have to argue that the agony of gathering people and a scuffle may be less than the reading on the meter. This may well be valid. But now comes the main difficulty. If that is so, why should the taxi driver not try the same tactic even if the passenger has paid? That is, he could take the fare and pretend that the passenger never paid and go through the same action as he would if the passenger had defrauded him, and thus end up collecting perhaps twice the correct fare, not to mention the tip. Everybody would agree, taxi drivers do not behave in this way. Therefore, they must be irrational, because it was supposed, a few lines ago that this behaviour is the one in conformity with their self-interest. (Some defenders of the faith would, however, be pleased to know that in the city of the author's residence, particularly in the late hours, taxi drivers do occasionally give evidence of rationality.)

Herein lies the crux of the matter. The object of the above exercise was not to show that human beings are not guided solely by selfishness, but to demonstrate that given the order that attends the multitude of economic exchanges in society and the absence of anarchy and fraud, this *must* be so. The 'invisible hand' would not be able to co-ordinate a multitude of selfish acts to bring order—as it is supposed to do—if it were not aided by the adherence of individuals to certain commonly accepted values. The example in this chapter shows that we can maintain that a subset of human beings conform to the 'law' entirely because of self-interest; but that rules out, *by implication*, the same assumptions for all the remaining individuals. Thus we have to make room for our sense of values, however small.

3.2

In a lot of economic theorizing it is *presumed* that all contracts are enforceable. Once this is granted, the efficiency of markets is ensured—barring of course the standard difficulties associated with externalities and returns to scale. It is only when considering markets like the one for loans, which are characterized by a long time-lag between the acts of the two parties involved in the exchange, do we talk of default (that is, the possibility of one party backsliding on his part of the contract). This is what has led to the substantial literature on credit market 'isolation' and 'interlinkage'. What is not always appreciated is that virtually all economic exchanges entail a time lag. Like that taxi driver, the barber brings the bill after the haircut, as does the waiter after the meal. And, as the above example shows, it is not possible to explain the absence of widespread default in these situations without making allowances for our sense of values and norms.

Thus while the absence of externalities, etc. is necessary for the efficiency of the invisible hand, a more *basic* assumption is that the agents involved in economic exchanges fulfil their obligations. And the ultimate guarantor of this assumption is our sense of values and norms. As Arrow puts it bluntly in his perceptive essay, 'A Cautious Case for Socialism' (1982), 'The model of *laissez-faire* world of total self-interest would not survive for ten minutes; its actual working depends on an intricate network of reciprocal obligations, even among competing firms and individuals'.

Once this is appreciated, it becomes possible to understand many features of society without recourse to artificial 'economic' arguments. Consider, for instance, the threat of violence. It is well known that one way in which a moneylender in a backward agricultural region ensures that money is repaid is by using the threat of violence. This would appear as a paradox to the economist ('Why does he not anyway use such a threat and extort money? Why does he bother to lend the money in the first place?'). But as soon as we accept the idea of norms and morality, such behaviour becomes easy to comprehend.

Similarly, to explain the larger incidence of default and fraud in economic transactions in some societies, we no longer need to claim an excess or a shortage of rationality on the part of their inhabitants, but may adduce the more reasonable explanation of differing social norms. And with this open up newer dimensions in policy making, in which social norms appear as a 'control variable'. What is important is to recognize that social norms can alter not only the society *but even the prices of goods and services.*

Finally, consider a variant of Sen's delightful application of the two-person game, the prisoner's dilemma, to a common social problem (Sen 1973). Let us assume, as is quite reasonable, that (i) every city dweller prefers his city to be clean rather than dirty, and (ii) one person throwing litter on the streets does not make a clean city dirty. It is easy to see that each individual, acting atomistically, would prefer to throw litter on the street rather than go through the trouble of looking for a garbage bin to dispose of it. It being rational for each individual to litter the streets, all citizens—if they were rational—would do so. The city would be a dirty one and [given (i)] everybody would be worse off.

I find this story convincing and therefore believe, though it sounds facetious, that the dirtiness of, for example, Calcutta, is a reflection of the rationality of its inhabitants. This also shows how much we can gain from a little bit of irrationality. Actually there are two ways of solving this problem. One is to impose fines for dirtying the streets; the other is to inculcate in human beings suitable values. The former works by changing what is rational to the individual. The latter works by making people accept a little bit of irrationality. It is true that the latter would take much longer to implement, but it is ethically clearly more attractive and ought to be the ultimate objective.

References

Arrow. 1982. 'A Cautious Case for Socialism'. In I. Howe, ed. *Beyond the Welfare State.* New York: Schocken Books.

Sen, A.K. 1973. 'Behaviour and the Concept of Preference'. *Economica* 40: 241–59

Part II
Games and Equilibrium Behaviour

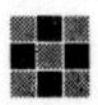

4 On the Non-existence of a Rationality Definition for Extensive Games

4.1. Introduction

In defining rationality in extensive games, unreached nodes cause some well-known problems.[1] In games of imperfect information one problem is that of finding suitable probability numbers for the nodes in the initial information set of subgames which are never actually reached. A more general problem which applies to games of both imperfect and perfect information is that standard solution concepts, like subgame perfection, implicitly require that players turn a blind eye to another player's 'irrationality' even if this has been revealed by virtue of having reached a node that could not have been reached had this player behaved rationally.

Attempts to solve this problem seem to run invariably into difficulties. The aim of the present chapter is to prove that the problem is, in fact, insoluble. There is a hint of this in some works (see for example, Binmore 1987). The aim of this chapter is to lend clarity to this debate by precipitating a formal impossibility theorem.[2] The theorem shows that a definition of rational behaviour which is applicable to all extensive games and which does not suffer from the problem of unreached nodes discussed above does not exist. This is because such a definition would run into difficulty with a class of repeated games which include the prisoner's dilemma and the games described by Rosenthal (1981) and Reny (1986).

From *International Journal of Game Theory* 19, 1990: 33–44.

I am grateful to T.C.A. Anant, Ken Binmore, Bhaskar Dutta, Vijay Krishna, Debraj Ray, Phil Reny, Ariel Rubinstein, Arunava Sen, and an anonymous referee for discussions and comments at various stages of the development of this paper. I also benefited from seminars at the Indian Statistical Institute, Delhi, and the London School of Economics.

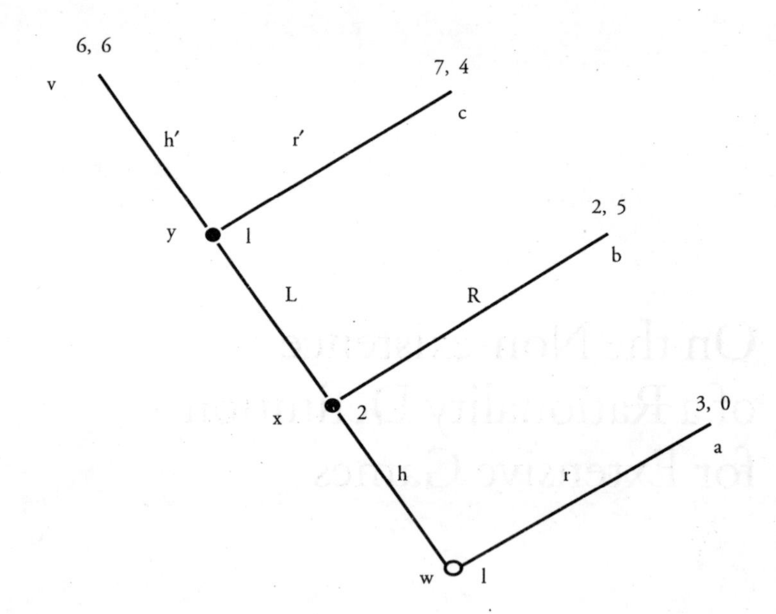

To motivate the problem consider the game Γ, described below, which is a truncated version of Rosenthal's (1981) game. The number at each node denotes the player who has to move there. The game is based on the following principle. At any node, if the relevant player moves left both players get 2 units each, if he moves right he gets 3 and the other gets zero. The pay-off at a terminal node is calculated by counting the number of left moves and right moves that the path entails.

By the backward-induction argument, it is clear that at each node, each player will want to move right, and hence the final outcome would be node a with a pay-off of (3,0). It is obvious that all standard solution concepts—Nash, subgame perfection, rationalizability—imply that the final outcome will be node *a*. But this seems intuitively unacceptable because introspection leads us to believe that, that is not the way we would play the game. This comes out more sharply if we consider a longer version of Γ in which there are, say, 100 non-terminal nodes. In such a game, if both players move left throughout they will get a pay-off of (200,200). It then seems very unlikely that players *will* opt to move right and end up with the pay-off (3,0). This is the same problem which arises in the finitely repeated Prisoner's Dilemma game. For the remaining discussion in this section it may be useful to think of the 100-move game instead of the 3-move one as in Γ.

This conflict between game-theoretic solution and intuition has provoked many attempts to explain formally why players would 'co-operate' (in the above example this entails moving left) in a game. An element of bounded rationality can explain why players may co-operate (Neyman 1985).[3] The same may be achieved by allowing for some uncertainty about the 'length' of the game (Basu

1987). A third and well-known avenue is to allow one player to entertain some doubt about the rationality of the other (Kreps et al. 1982).

There is, however, a more fundamental difficulty with all these escape routes. Suppose two experts in game theory—Messrs. von Neumann and Morgenstern for example—were made to play the prisoner's dilemma one hundred times. Most of us would maintain that even here there will not be defection in all hundred games.[4] Here both players are rational and the possibility of one player feeling that the other is irrational is too remote to be treated as an explanation of why co-operation may occur.

In an earlier paper (Basu 1988), I tried to develop a solution concept based on the argument that if in the first move player 1 co-operates, player 2 will see that a node has been reached which could not have been reached if 1 was 'rational'. Once 2 doubts 1's rationality, it may be reasonable for 2 to play co-operatively (at least for some time). But surely 1 can see this and so 1 may actually play co-operatively in the first game. In other words, he plays an 'irrational' move in order to confuse 2. His irrationality is a strategic one.

An important objection can be raised against this. Since 1 manages to do well by feigning 'irrationality', it can be argued that this ought not to be described as 'irrationality' at all.[5] And since others can see this, they will not be misled into believing that 1 is irrational.

It seems to me, however, that the onus of averting this problem lies on the rationality definition being used. This is the line taken in the present chapter. The *rationality* definition must be such as to render strategic irrationality ineffective. Once this requirement is coupled with a weak backward-induction axiom we find that it is not just subgame perfection and rationalizability which run into difficulty but so do all existing and potential solution concepts. This is formally established in Section 4.2, and Sections 4.3 and 4.4 discuss and interpret the axioms and the theorem.

4.2. The Impossibility Theorem

The term *game* will be used here to refer to any perfect information, no nature-move, two-player, extensive-form game in the sense of Selten (1975) or Kreps and Wilson (1982). A brief resume of the notation follows; for elaboration the reader is referred to Kreps and Wilson.

Given a node, x, $\alpha(x)$ will be used to denote the *last action* taken to reach x and $p(x)$ to denote the node *immediately preceding* x. In the game described above, $\alpha(y) = L$, $\alpha(b) = R$, and $p(y) = p(b) = x$. Player i's utility function (defined on the terminal nodes) is denoted by u^i. Given a game, Γ, the subgame which consists of node x and all its successors is denoted by Γ_x.

A strategy of player i is a mapping s^i which specifies an action at every node where i is supposed to play. S^i is the set of all strategies of i.

A solution concept, R, is any mapping defined on the domain of all games such that, given any game Γ, it specifies a non-empty collection of strategy-pairs (one for each player) in the game. Hence $\phi \neq R(\Gamma) \subset S_1 \times S_2$ where S_i is the set of strategies of i in the game Γ. We refer to $R(\Gamma)$ as the *solution* of Γ (under R).

In the context of individual-rationality discussions, it is convenient to impose a technical assumption, which I shall call the *factorability* axiom.

AXIOM F: The solution concept, R, must be such that for all Γ, if $(s_1, s_2) \in R(\Gamma)$ and $(\hat{s}_1, \hat{s}_2) \in R(\Gamma)$, then $(s_1, \hat{s}_2) \in R(\Gamma)$.

Axiom F suggests that whether i's strategy s_i is rational or not is independent of what others are using. There are good reasons for defending axiom F (see Bernheim 1984; and Pearce 1984), but there is no need to enter this controversy here because in the present context it is easily seen to be irrelevant by taking any of the two following approaches. We could in this chapter restrict attention to games in which for each player i the terminal nodes are strictly ordered. That is, $x \neq y$ implies $u_i(x) \neq u_i(y)$. Since our aim is to prove impossibility, nothing is lost by such domain restriction. Secondly, we could interpret standard solution concepts in a way that is consonant with definition 4.1. If $\hat{R}$ is the Nash solution, we could think of R as a counterpart of this if R is such that, for all Γ, $R_i(\Gamma)$ is the set of all $s_i \in S_i$ such that there exists a strategy pair $(s_1{}^i, s_2{}^i) \in \hat{R}(\Gamma)$ where $s_i^i = s_i$.

A solution concept embodies our notion of rationality. If in a game a node is reached that cannot be reached if all players employ strategies in the solution set, it means some player has behaved 'irrationally'. Hence, in any game, Γ, given a solution concept, R, we can at each node, x, specify the set of players, $\Omega^{R\Gamma}(x)$, who have been revealed irrational. The method of doing this is outlined in definition 4.1. [We shall assume throughout that the solution concept, R, satisfies axiom F. This allows us to write $R(\Gamma) = (R_1(\Gamma), R_2(\Gamma))$.]

DEFINITION 4.1: Given a solution concept, R, and a game, Γ, we define an irrationality map

$$\Omega^{R\Gamma}: T \rightarrow \{\phi, \{1\}, \{2\}, \{1,2\}\}$$

(where T is the set of all nodes in Γ) by induction as follows:

$$\Omega^{R\Gamma}(w) = \phi$$

(where w is the initial node) and for all non-initial nodes x, we have

$\Omega^{R\Gamma}(x) = \{i \mid i \in \Omega^{R\Gamma}(p(x))$ or at $p(x)$ it was i's move and there does not exist $s \in R_i(\Gamma)$ such that $s(p(x)) = \alpha(x)\}$.

Definition 4.1 tells us that, as in conventional game theory, all players are assumed to be rational to start with (that is, the set of irrational players, $\Omega^{R\Gamma}(w)$, is empty). At any other node x a player is irrational [that is, he belongs to the set $\Omega^{R\Gamma}(x)$] if either he is already revealed irrational or at the previous node it was his move and there is no rational strategy of his which could have made him choose the alternative that brings us to node x.

As soon as a solution concept is specified, an implicit rationality definition gets specified as well. It is worth stressing that $R_i(\Gamma)$ is not a definition of rationality but it *embodies* a definition of rationality. The best way to think of a rationality definition is as a statement of moves that a rational player may make at each node. Thus if x is a node in Γ where i has to move and $i \notin \Omega^{R\Gamma}(x)$ then all moves m such that there exists $s \in R_i(\Gamma)$ for which $s(x) = m$ are rational moves. If $i \in \Omega^{R\Gamma}(x)$ then i's move at x, of course, does not tell us what a rational player would do.

I shall first develop a very weak form of the backward-induction axiom. For this we need some new notation. For any strategy n-tuple s in the game Γ, let $s : \Gamma x$ represent the restriction of s on the subgame Γx. We shall use $\theta (R, \Gamma, x)$ to represent the set of terminal nodes which can be reached in Γ_x by $s : \Gamma_x$ for some $s \in R(\Gamma)$.

If Z_1 and Z_2 are subsets of the set of terminal nodes Z in some game Γ, we write $Z_1 >_i Z_2$ if for all $x \in Z_1$ and for all $y \in Z_2$, $u_i(x) > u_i(y)$.

Our backward-induction axiom asserts the following: suppose in some game y and v are immediate successors of x and it is i's move at x. If all the terminal nodes that can be reached from v by playing strategies in the solution set dominate from i's point of view all the terminal nodes that can be reached from y by playing strategies in the solution set, then if i moves so as to get to y then i is revealed irrational.

Axiom B: The solution concept, R, must be such that for all Γ and for all nodes x, y, v, where y and v are immediate successors of x, if it is i's move at x and $\theta (R, \Gamma, v) >_i \theta (R, \Gamma, y)$, then $i \in \Omega^{R\Gamma}(y)$.

In Basu (1988), co-operation in the prisoner's dilemma is explained by allowing a player to make a move that shows him up as irrational and this influences the play of others in a way that could be beneficial to the player. A reasonable objection to this is that other players can surely see through such *strategic* irrationality and hence would not view the original player as irrational at all. To state this formally, I shall abuse the $\theta (\cdot)$ notation developed above a little and use $\theta (s_j, \Gamma, x)$ to denote the set of terminal nodes which can be reached in Γ_x by $s : \Gamma_x$ where s is any strategy pair whose j-th component is s_j.

The next axiom states that if in some game, y and v are immediate successors of x and it is i's move at x and for some strategy of the other player belonging to the solution set, all the terminal nodes reachable from y dominate (from i's point of view) all the terminal nodes reachable from v, then if i is described as rational at v, i must be described as rational at y.

Axiom S: The solution concept, R, must be such that for all Γ and for all nodes x, y, v where y and v are immediate successors of x, if there exists $s_j \in R_j(\Gamma)$ such that $\theta (s_j, \Gamma, y) >_i \theta (s_j, \Gamma, v)$ and $i \notin \Omega^{R\Gamma}(v)$, then $i \notin \Omega^{R\Gamma}(y)$.

Before stating the next axiom, note that in the approach taken in this chapter (in contrast to the traditional, extensive game model) $R_i(\Gamma_x)$ need not be equal to $R_i(\Gamma): \Gamma_x$. That is, the set of i's possible strategies in the game Γ_x need not coincide with the restriction of i's possible strategies in Γ to the subgame Γ_x. This is because Γ_x considered as a game in itself implies that no one is irrational and i knows this. But in Γ when node x is reached, some players may have been revealed irrational. Even though i himself may not have been revealed irrational, his play in the subgame Γ_x may be influenced by his awareness that there are players who are known to be irrational. It is in this sense that my approach may be described as history-sensitive. How a player plays in a subgame depends on the history of play at the initial node of the subgame.

Since my argument turns on the definition of rationality, it is important to specify what irrationality implies. A simple assumption is that an irrational player is unpredictable.

In terms of the game, Γ, described above, if 1 is known to be irrational at node y, then 1 (it is expected) may play h' or r'.

AXIOM U: The solution concept, R, must be such that for all Γ and for any node x if $i \in \Omega^{RT}(x)$ and it is i's move at x, then $R_i(\Gamma) : \Gamma_x = S_i : \Gamma_x$, where S_i is player i's set of strategies in the game Γ.

The axiom simply states that once a player has been revealed irrational, from there onwards he is treated as unpredictable.

Axioms F, B, S, and U are together incompatible.

THEOREM 4.1: There does not exist any solution concept satisfying axioms F, B, S, and U.

PROOF: This theorem is weaker than Theorem 4.2 proved below.

All these axioms can be relaxed without losing the impossibility theorem. This is discussed in a later section but one particular axiom weakening is worth discussing here. It may seem to some that an irrational player should not be treated as totally unpredictable. In that case we may wish to weaken axiom U to state the following. An irrational player is less predictable than a rational player. That is, he may play any strategy that a rational player may play and (wherever possible) there are other strategies which he may play. We no longer require that he may play *any* strategy in his strategy set as in axiom U.

AXIOM U*: The solution concept, R, must be such that for all Γ and for any node x, if it is i's move at x, $i \in \Omega^{RT}(x)$ and $R_i(\Gamma_x)$ is a proper subset of $S_i : \Gamma x$, then $R_i(\Gamma_x)$ is a proper subset of $R_i(\Gamma): \Gamma_x$.

The axiom says that if at a node x, i is considered irrational, then the set of strategies that he may be expected to employ thereon (that is, $R_i(\Gamma): \Gamma_x$) is a proper super set of the strategies that a rational player may be expected to employ in a similar situation [i.e., $R_i(\Gamma_x)$], assuming, of course, that $R_i(\Gamma_x)$ is not already as large as is feasible (i.e., $R_i(\Gamma_x)$ is not equal to $S_i : \Gamma_x$).

To illustrate this axiom suppose Γ, shown above, is actually a subgame of a larger game. In this subgame player 1 has 4 strategies: rr' rh', hr', hh'. Suppose that if 1 is rational, our solution predicts I will play rr'. What axiom U* says is that if, at node w, 1 is known to be irrational, the solution must predict a larger set of possible strategies that 1 may employ in this subgame. That is, the predicted set of strategies must include rr' and one or more strategies from the three remaining available.

Actually we could think of an intuitively more correct version of U*. Note that the subgame Γ_x considered as a game in itself treats all players (not just i) as rational. Hence the difference between $R_i(\Gamma)$: Γ_x and $R_i(\Gamma_x)$ in axiom U* is not just that the former takes into account i's irrationality and the latter does not. Instead, the latter treats *everyone* as rational. If we want the only distinction to be i's rationality, then we would have to write axiom U* more elaborately as follows.

The solution concept, R, must satisfy the following condition: for all Γ and for all $x \in T$, if it is i's move at x, $i \in \Omega^{R\Gamma}(x)$ and there exists a game $\hat{\Gamma}$ such that $\hat{\Gamma}_y = \Gamma_x$ and $\Omega^{R\Gamma}(x) - \Omega^{R\hat{\Gamma}}(y) = \{i\}$ and $R_i(\hat{\Gamma}) : \hat{\Gamma}_y$ is a proper subset of $S_i : \hat{\Gamma}_y$ then $R_i(\hat{\Gamma}) : \hat{\Gamma}_y$ is a proper subset of $R_i(\Gamma) : \Gamma_x$.

It will be clear from the proof of Theorem 4.2 that it does not matter formally whether we use axiom U* or its more intuitively appealing variant just described above. Hence, I use the technically simpler axiom U*.

THEOREM 4.2: There does not exist any solution concept satisfying axioms F, B, S, and U*.

PROOF: Let R be a solution concept that satisfies axioms F, B, S, and U*. Axiom F allows us to speak of R_1 (Γ) and R_2 (Γ) independently. Consider the game, Γ, described in Section 4.1. It follows from definition 4.2 that either of the following must be true: $\Omega^{R\Gamma}(x) = \phi$ or $\Omega^{R\Gamma}(x) = \{1\}$.

Suppose $\Omega^{R\Gamma}(x) = \phi$. Then $1 \notin \Omega^{R\Gamma}(y)$. By axiom B we know $1 \in \Omega^{R\Gamma}(v)$. Let $s \in R_1$ (Γ). Since $1 \notin \Omega^{R\Gamma}(y)$ and $1 \in \Omega^{R\Gamma}(v)$, definition 4.1 implies $s(y) \neq h'$. Hence, $s(y) = r'$.

This implies $\theta\,(R, \Gamma, b) >_2 \theta\,(R, \Gamma, y)$. Hence, by axiom B, $2 \in \Omega^{R\Gamma}(y)$. Since $2 \notin \Omega^{R\Gamma}(x)$, definition 4.1 implies that if $s' \notin R_2$ (Γ), then $s'(x) = R$.

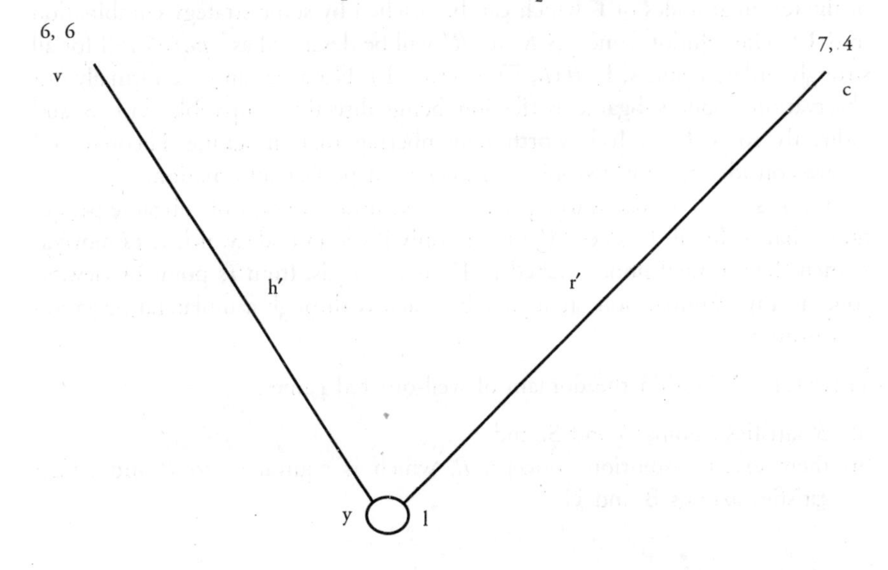

By repeating the same argument at w, it can be shown that if $s \in R_1(\Gamma)$, then $s(w) \neq h$. Hence, $1 \in \Omega^{R\Gamma}(x)$, which is a contradiction.

Suppose $\Omega^{R\Gamma}(x) = \{1\}$. It follows from definition 4.1 that $1 \in \Omega^{R\Gamma}(y)$. Consider now the game Γ_y described below:

By axiom B we know that if $s \in R_1(\Gamma_y)$ then $s(y) = r'$. Hence $R_1(\Gamma_y)$ is a proper subset of $S_1: \Gamma_y$. Since player 1 has to play at y and $1 \in \Omega^{R\Gamma}(y)$, we know by axiom U* that $R_1(\Gamma_y)$ must be a proper subset of $R_1(\Gamma) : \Gamma_y$. Hence the strategy s_1 of player 1 such that $s_1(y) = h'$ must be an element of $R_1(\Gamma) : \Gamma y$. Hence there exists $s_1 \in R_1(\Gamma)$ such that $v \in \theta(s_1, \Gamma, y)$. Hence $\theta(s_1, \Gamma, y) >_2 \theta(s_1, \Gamma, b)$. Axiom S implies that if $2 \notin \Omega^{R\Gamma}(b)$, then $2 \notin \Omega^{R\Gamma}(y)$. Hence, if $2 \in \Omega^{R\Gamma}(y)$, then $2 \in \Omega^{R\Gamma}(b)$. But this is impossible by definition 4.1, since $2 \notin \Omega^{R\Gamma}(x)$. Hence $2 \notin \Omega^{R\Gamma}(y)$. By definition 4.1, there exists $s_2 \in R_2(\Gamma)$ such that $s_2(x) = L$. Hence, $\theta(s_2, \Gamma, x) >_1 \theta(s_2, \Gamma, a)$. By axiom S, $1 \notin \Omega^{R\Gamma}(x)$, which is a contradiction.

4.3. Discussion

How does subgame perfection relate to my axioms? It will be shown here that none of my axioms, B, S, and U, is *individually* incompatible with subgame perfection (though, of course, taken together, they are). While subgame perfection directly satisfies B and S, its compatibility with U is more indirect and in a sense that needs to be made precise. This is done in the next theorem, but before stating the theorem we need one more definition.

In order to refrain from the technical, but conceptually inconsequential difficulties that could arise from the fact that a solution concept does not automatically satisfy axiom F, I shall here restrict attention to games in which each player has a strict ordering of the terminal nodes. Such games will be referred to as *strongly ordered games*.

Abusing the $\theta(\cdot)$ notation once more, let us use $\theta(R, \Gamma)$ to denote the subset of the terminal nodes of Γ which can be reached by some strategy combination in R(Γ). The solution concepts R and R' will be described as *equivalent* if for all strongly ordered games, Γ, $\theta(R, \Gamma) = \theta(R', \Gamma)$. Now we can state formally our observation about subgame perfection being directly compatible with S and indirectly so with U. It is worth remembering that in stating Theorem 4.3 we are considering only two-player games with perfect information.

In this section we use $\hat{R}$ to represent the solution concept of subgame perfection. That is, for all Γ, $s^* \in \hat{R}(\Gamma)$ if and only if for all node x, if it is i's move at x, then the terminal node reached in Γ_x by $s^* : \Gamma_x$ is, from i's point of view, as good as any terminal node that may be reached through a unilateral deviation by i from s^*.

THEOREM 4.3: Within the domain of well-ordered games,

(i) $\hat{R}$ satisfies axioms B and S, and
(ii) there exists a solution concept, R, which is equivalent to $\hat{R}$ and which satisfies axioms B and U.

PROOF: (i) Consider a strongly ordered game Γ and nodes x, y, and v, where it is i's move at x, y and v are immediate successors of x, and $\theta(\hat{R}, \Gamma, y) >_i \theta(\hat{R}, \Gamma, v)$.

Suppose $i \notin \Omega^{\hat{R}\Gamma}(v)$. Hence, there exists $s \in \hat{R}_1(\Gamma)$ such that $s(x) = \alpha(v)$. But it is easy to see that there exists $s' \in S_i$ which would in the subgame Γ_x take i to a superior terminal node than s. For this s' must simply be such that $s'(x) = \alpha(y)$. Hence $s \notin \hat{R}_1(\Gamma)$.

This contradiction establishes that $i \in \Omega^{\hat{R}\Gamma}(v)$. Hence $\hat{R}$ satisfies axiom B. A similar proof can be constructed to show that $\hat{R}$ satisfies S.

(ii) Let us construct a solution concept, R, as follows. For all strongly ordered games, Γ, and for all players, i, in Γ, $R_i(\Gamma)$ is defined in the following way: $s \in R_i(\Gamma)$ if and only if there exists $s' \in \hat{R}_i(\Gamma)$ such that for all x where i has to move and $i \notin \Omega^{\hat{R}\Gamma}(x)$, $s(x) = s'(x)$. It is now easy to check that R is equivalent to $\hat{R}$ and it satisfies axioms B and U.

4.4. INTERPRETATION

The discussion in the previous section suggests that an attempt to break out of the impossibility established in Section 4.2 should be focused on axiom U or U*. Before going into a discussion of the intuition for accepting or rejecting this axiom, I want to demonstrate that the above theorem can be strengthened by weakening U or U* in an interesting way. For linguistic simplicity, I shall conduct the discussion here in terms of U, though much the same could have been said using U*. Note that what axiom U asserts is that a player who has once been observed behaving irrationally must be, then onwards, treated as completely unpredictable. I feel this is a better assumption than the traditional one which would ignore the revealed irrationality and continue to treat the player as completely rational. It may legitimately be argued, however, that these two polar assumptions are not the only possible ones; and there may be an escape route in-between. The aim of this section is to show that these 'in-between' routes are all blocked because Theorems 4.1 and 4.2 can be generalized to more powerful impossibility theorems.

Let us consider the following weakening of U. Suppose R is the solution concept, and let Γ be a game in which at node x it becomes clear that i is not playing any strategy in $R_i(\Gamma)$. Axiom U would abandon attempts to predict i's play here onwards. However, we could think of an intermediate approach where we have a predefined (first) fall-back rationality definition, R_i^1, such that $R_i^1(\Gamma)$ is a proper subset[6] of S_i, and a player i who violates the rationality definition R_i is expected to play some strategy from the set $R_i^1(\Gamma)$. In other words, $R_i^1(\Gamma)$ is like a second hypothesis about what player i may do, once the first hypothesis [embodied in $R_i(\Gamma)$] is proved false.

At first sight, it seems that modifying axiom U in this manner is a way out. But it is not. This is because it is always possible to construct a game tree, a little more elaborate than the one used in the proof of Theorem 4.2, in which there is a node which has the following history. It requires that i violates rationality definition, R_i, and then violates the (first) fall-back rationality definition, R_i^1. What should we assume about *i*'s play here onwards? If we now abandon attempts to predict *i*'s play and treat him as completely unpredictable, then an impossibility result can be proved as before. But, of course, there is a way out. This is, to assume that we have a second fall-back rationality definition, R^2, for a player who first violates R^2 and then violates R^1.

But by now it should be evident to the reader that as long as we have a *finite* number of fall-back rationality definitions and concede that a player who has violated all these must be totally unpredicable, we can combine this weaker version of axiom U, with axioms F, B, and S to once again establish an impossibility theorem.

Despite the reliance on one principal game in the proofs, it should be obvious that the paradox highlighted in this chapter arises in a class of games, including Selten's (1978) chain-store game, the prisoner's dilemma, Reny's (1986) game, and combinations of these. It should, in principle, be possible to isolate the class of games in which the rationality paradox arises. I did not attempt this in the present chapter but it is worth noting that, in a broad sense, the paradox arises in any game which has the following possibility: There is a move such that after a player (call him A) makes the move then (i) if others treat A as irrational, it influences their future play in a way which makes the move a rational one and (ii) if others treat A as rational, then it influences their future play in a way which makes the move irrational for A.

It should be immediately clear that if a certain move is the last one A makes in a game, then that move can never be the basis of the above conflict. It follows, that in repeated games with *changing* partners (as in some biological games) the kind of paradox precipitated in this chapter does not arise. Subgame perfection may, therefore, be a solution concept more suitable for evolutionary games.

Let us now turn to some interpretational issues. As already mentioned, the axiom that is likely to be the most contentious is axiom U or one of its variants. Hence, in the light of the above impossibility theorems, we are forced into taking one of the two broad positions. (1) We could treat U or U* as reasonable and reject the view that every game must have a solution, or that rationality is always definable even in strategic environments. (2) We could reject U and U* and maintain that reasonable solutions can be defined for all games.

(2) seems to direct us towards a model with 'mistakes'. The usual approach is to consider the limit of games with mistakes (as the mistakes vanish). An alternative is a model in which deviations from rationality are treated as possible but of zero-probability, in a measure-theoretic sense. This makes it possible to ignore past deviations. This is an interesting route which has not been adequately explored in the literature.[7]

However, as must be evident from the many arguments above, I am at this point inclined to go along with (1). Hence, the present chapter considered only a 'no-mistakes' framework. This is not unreasonable since in games in which the *rules* of play are simple, (for example, the prisoner's dilemma in contrast to chess) the scope for mistakes is indeed quite limited. In the light of these comments, one way of interpreting the theorems in this chapter is to treat them as suggesting that rationality cannot be defined unless allowance is made for mistakes. It is not that this chapter says that if such an allowance is made, we can define rationality; but it says that the other route is certainly closed.

NOTES

1. To cite a few references from a large literature: Rosenthal (1981); Kreps and Wilson (1982); Binmore (1987); Reny (1986); Cho (1987); Kreps and Ramey (1987); Dekel and Fudenberg (1987); and Basu (1988).
2. Abreu and Pearce (1984) have also established some impossibility results for solution concepts, but their axiom structure and motivation are quite different from the ones used here.
3. It is worth noting, though, that Rubinstein's (1986) model, using a similar concept of bounded rationality, rules out co-operation in the prisoner's dilemma.
4. An earlier version of this chapter figured two *contemporary* game-theorists, but I had to change the names after Ken Binmore assured me, to my dismay, that one of them would defect each time.
5. Amartya Sen took such a line in his Yrjo Jahnsson lectures in Helsinki in 1987.
6. If $R_i^1(\Gamma) = S_i$, then we are back to the *complete* unpredictability assumption of Section 4.2.
7. I owe this suggestion to an anonymous referee.

REFERENCES

Abreu, D., and D.G. Pearce. 1984. 'On the Inconsistency of Certain Axioms on Solution Concepts for Non-Cooperative Games'. *Journal of Economic Theory* 34: 169–74.

Basu, K. 1987. 'Modeling Finitely-Repeated Games with Uncertain Termination'. *Economics Letters* 23: 147–51.

——. 1988. 'Strategic Irrationality in Extensive Games'. *Mathematical Social Sciences* 15: 247–60.

Bernheim, D. 1984. 'Rationalizable Strategic Behaviour'. *Econometrica* 52: 1007–28.

Binmore, K. 1987. 'Modelling Rational Players'. *Economics and Philosophy* 3: 179–214.

Cho, I. 1987. 'A Refinement of Sequential Equilibrium'. *Econometrica* 55: 1367–89.

Dekel, E., and D. Fudenberg 1987. 'Rational Behaviour with Payoff Uncertainty'. Cambridge, Massachusetts, Working Paper.

Kreps, D.M., P. Milgrom, and R. Wilson. 1982. 'Rational Cooperation in the Finitely-repeated Prisoner's Dilemma'. *Journal of Economic Theory* 27: 245–52.

Kreps, D.M and G. Ramey. 1987. 'Structural Consistency, Consistency, and Sequential Rationality'. *Econometrica* 55: 1331–48.

Kreps, D.M. and R. Wilson. 1982. 'Sequential Equilibria'. *Econometrica* 50: 863–94.

Neyman, A. 1985. 'Bounded Complexity Justifies Cooperation in the Finitely Repeated Prisoner's Dilemma'. *Economics Letters* 19: 227–9.

Pearce, D.G. 1984. 'Rationalizable Strategic Behaviour and the Problem of Perfection'. *Econometrica* 52: 1029–50.

Reny, P. 1986. 'Rationality, Common Knowledge and the Theory of Games'. PhD dissertation, Princeton University.

Rosenthal, R.W. 1981. 'Games of Perfect Information, Predatory Pricing and the Chain Store Paradox'. *Journal of Economic Theory* 25: 92–100.

Rubinstein, A. 1986. 'Finite Automata Play the Repeated Prisoner's Dilemma'. *Journal of Economic Theory* 39: 83–96.

Selten, R. 1975. 'Reexamination of the Perfectness Concept for Equilibrium Points in Extensive Games'. *International Journal of Game Theory* 4: 25–55.

——. 1978. 'The Chain-store Paradox'. *Theory and Decision* 9: 127–59.

5 Strategy Subsets Closed under Rational Behaviour

with Jorgen W. Weibull

5.1. Introduction

While a Nash equilibrium is a point in the Cartesian product of the players' strategy spaces such that no player can increase his pay-off by a unilateral deviation, a strict Nash equilibrium requires that any unilateral deviation actually incurs a loss. Any strict Nash equilibrium has all the strategic stability properties that the refinement literature asks for, but many games lack such equilibria. In any non-strict Nash equilibrium, at least one player is indifferent between some of his pure strategies even under his Nash-equilibrium beliefs. As is well-known, such indifference can make the equilibrium highly 'unstable'.

The point is illustrated in Figure 5.1. The game in diagram (a) has a unique Nash equilibrium, and in this equilibrium both players randomize between their first two strategies, player 1 choosing T with probability 2/3 and M with probability 1/3 and player 2 choosing L with probability 1/4. However, under these Nash equilibrium beliefs, player 1 is indifferent between T and M, and player 2 is indifferent between L and R. If player 1 would assign probability $p >$ 2/3 to the uncertain event that 2 will choose L, then 1's unique optimal strategy is B. And if player 2 would assign probability $q > 1/3$ to the event that 1 will choose M, then 2's unique strategy is L etc. Strategy T, which is assigned probability 2/3 in Nashs equilibrium, is optimal for player 1 only if he assesses $1/4 < p < 1.3$. It thus appears that the unique Nash equilibrium of this game (hence its only perfect equilibrium) is 'unstable'. In particular, it does not seem

From *Economics Letters* 36, 1991: 141–6.

We are grateful for helpful comments from Dilip Abreu and Brian Kanaga. The research of Weibull was supported by the Swedish Council for Research in the Humanities and Social Sciences (HSFR).

	L	R
T	4,1	1,2
M	1,3	2,1
B	6,2	0,3

(a)

	L	R
T	2,2	0,0
M	0,1	1,1
B	0,0	0,2

(b)

Figure 5.1

justified to exclude, as does the Nash equilibrium criterion, the possibility that player 1 will play *B*, even if pre-play communication is presumed.

While the example in Figure 5.1(a) suggests a coarsening of the Nash equilibrium concept, the same type of argument applied to the game in Figure 5.1(b) suggests a refinement of the Nash-equilibrium requirement. For although the Nash equilibrium (*M, R*) is undominated and hence perfect, a pre-play agreement to play (*M, R*) does not appear to be 'self-enforcing', since also *L* is a best reply to *M*, and in view of this indifference on behalf of player 2, player 1 may contemplate playing *T*. If player 2 assigns positive probability to the possibility that 1 will play *T*, and a smaller probability to the event that 1 will play B, then 2's unique optimal strategy is *L*, in which case player 1 should certainly play *T* instead of *M*. Hence, it seems that we are led to a rejection of the perfect but non-strict Nash equilibrium (*M, R*) in favour of the strict equilibrium (*T, L*).

We will argue that even if one agrees to treat pre-play communication as possible but implicit, it is not clear why one should presume that 'agreements' take the form of a single strategy (pure or mixed) profile, rather than a set of strategy profiles. Indeed, a set-valued solution concept has been developed by Kohlberg and Mertens (1986). However, while they select sets of Nash equilibria, that is, sets contained in their own best replies, we here select sets containing all their own best replies, a 'dual' approach which can be viewed as a set-theoretic coarsening of the notion of strict Nash equilibrium while their approach is a set-theoretic refinement of the Nash equilibrium concept.

One set-valued solution concept developed in this chapter is the 'tight curb' notion, that is, sets which are identical with their own best replies. In the light of the work of Bernheim (1984) and Pearce (1984) one sees that, in a game with continuous pay-off functions and compact strategy sets, a maximal tight curb set coincides with the set of rationalizable strategy profiles. Hence, while their work may be seen as an investigation into the properties of the maximal tight curb set in a game, the present chapter may be viewed as an exploration of the whole spectrum of tight curb sets, with particular attention paid to the opposite end of this spectrum, viz., the minimal tight curb sets. One reason for highlighting minimal tight curb sets is that these are the 'nearest' set-valued generalizations of strict equilibria. Moreover, minimality has the advantage of reducing strategic ambiguity. However, in certain games such an advantage turns out to be too costly in terms of other game-theoretic desiderate, so in this first exploration we consider minimal tight curb sets only as useful benchmarks. In addition to studying curb and tight curb sets, defined in terms of best replies, we also investigate a weaker notion, *curb** sets, based on undominated best replies.

5.2. Notation and Preliminaries

Attention in this chapter is focused on normal-form games $G = (N, S, U)$, where $N = \{1, 2,..., n\}$ is the set of players, S is the Cartesian product of the players' strategy sets S_i, and U is a mapping of S into $\mathbb{R}^n$, such that $U_i(s) \in \mathbb{R}$ is the i'th player's von Neumann-Morgenstern utility level when strategy profile s is played. We will let $\mathcal{G}$ denote the class of all games G in which each strategy set S_i is a compact set in some Euclidean space and each pay-off function U_i: $S \rightarrow \mathbb{R}$ is continuous. For any game G, let P be the collection of all products of non-empty and compact subsets of the players' strategy sets, that is, $X \in P$ if and only if X is the Cartesian product of non-empty compact sets $X_i \subset S_i$ $[i = 1, 2,...,n]$. (In particular, $S \in P$ if $G \in \mathcal{G}$.) As is usual, we will write $s \setminus s'_i$ when the ith component of a profile s is replaced by s'_i. A strategy profile $s \in S$ is a Nash equilibrium if $U_i(s) \geq U_i(s \setminus s_i') \; \forall i \in \mathbb{N}, \forall s_i' \in S_i$, and it is a strict (Nash) equilibrium if $U_i(s) > U_i(s \setminus s_i') \; \forall i \in \mathbb{N}, \forall s_i' \in S_i - \{s_i\}$.

In the present chapter, a player's belief about others' strategies takes the form of a product-probability measure on all players' strategy sets (we include his own strategy set in the domain only for notational convenience). Hence, a player's belief is formally identical with a mixed strategy profile $m \in M(S)$, where $M(S)$ is the Cartesian product of the sets $M(S_i)$ of Borel probability measures over each strategy set S_i. For any Borel subset $X_i \subset S_i$, let $M(X_i)$ be the (Borel) probability measures with support in X_i, i.e., $M(X_i) = \{m \in M(S_i): m(X_i) = 1\}$, and for any $X \in P$ let $M(X)$ be the Cartesian product of the sets $M(X_i)$.

For each player $i \in \mathbb{N}$, strategy $s_i \in S_i$ and belief $m \in M(s)$, let $u_i(s_i \mid m)$ be the player's expected utility under belief m when he plays s_i.[1] Let $\beta_i(m)$ be the i'th player's set of optimal strategies in S_i under belief $m \in M(S)$, i.e., strategies $s_i \in S_i$ such that $u_i(s_i \mid m) > u_i(s_i' \mid m) \; \forall s'_i \in S_i$. If $G \in \mathcal{G}$, that is, pay-off functions are continuous and strategy sets compact, then each set $\beta_i(m) \subset S_i$ is non-empty and compact. For any set $X \in P$, let $\beta_i(X)$ denote the i'th player's set of optimal strategies under beliefs in $M(X)$:

$$(5.1) \qquad \beta_i(X) = \bigcup_{m \in M(X)} b_i(m),$$

and write $\beta(X)$ for the Cartesian product of the sets $\beta_i(X)$. This study is restricted to games G in $\mathcal{G}$, a class of games for which $X \in P$ implies $\beta(X) \in P$.

5.3. Sets Closed Under Rational Behaviour

A set X of strategy profiles will be said to be *closed under rational behaviour* (*curb*) If $X \in P$ and $\beta(X) \subset X$. In words: a set X in P is curb if the belief that strategies outside X will not be played implies that such strategies will indeed not be played by players who are rational in the sense of never playing strategies that are sub-optimal.

In spirit, the present criterion is related to the notion of strict equilibrium, and, indeed, every such equilibrium, viewed as a singleton set, meets the curb condition: if $s \in S$ is a strict equilibrium, then $\{s^*\} \subset \beta(\{s^*\})$. However, the curb criterion is also met by the set $X = S$ of all strategy profiles in the game, the set $X = S$ thus being the maximal curb set. Conversely, one may ask whether there exist *minimal* curb sets, that is, curb sets which do not contain any proper subset which is a curb set.

PROPOSITION 5.1: Every game $G \in \mathscr{G}$ possesses at least one minimal curb set.

PROOF: Let Q be the (non-empty) collection of curb sets in S, partially ordered by (weak) set inclusion. By Hausdorff's Maximality Principle, Q contains a maximal nested sub-collection. Let $Q' \subset Q$ be such a sub-collection, and, for each $i \in$ N, let $\tilde{X}_i$ be the intersection of all sets X'_i for which $X' \in Q'$. Since each set X'_i is non-empty and compact, so is $\tilde{X}_i$, by the Cantor Intersection Theorem. Hence, $\tilde{X} \in P$. Suppose $s_i \in \beta_i(\tilde{X})$. Since $M(\tilde{X}) \subset M(X') \forall X' \in Q'$, we have $s_i \in \beta_i(X') \ \forall X' \in Q'$, and thus $s_i \in X_i' \ \forall X' \in Q'$ (since all $X' \in Q'$ are curb). Hence, $s_i \in \tilde{X}_i$, so $\beta_i(\tilde{X}) \subset \tilde{X}_i \ \forall_i \in$ N, that is, $\tilde{X}$ is curb.

Generalizing the definition of strict equilibrium from singleton sets to arbitrary product sets, we call a curb set X tight if $\beta(X) = X$.[2] In particular, a profile $s \in S$ is a strict equilibrium if and only if {s} is a tight curb set. Note that a tight curb set is 'immune' to iterated elimination of sub-optimal strategies under beliefs in $M(X)$. For if it is common knowledge that no player will use a strategy outside X, then each player knows that other (rational) players will play in $\beta(X)$ and hence each player should play in $\beta(\beta(X))$ etc. If X is tight curb, then such iteration has no effect: $\beta^n(X) = X$, for all n.

While many games lack strict equilibria, every game G in $\mathscr{G}$ possesses at least one tight curb set. In fact, in such games, every minimal curb set is tight, indeed a minimal tight curb set (that is, it contains no proper subset which is a tight curb set). Conversely, every minimal tight curb set is a minimal curb set. Formally:

PROPOSITION 5.2. A set in a game $G \in \mathscr{G}$ is a minimal curb set if it is a minimal tight curb set.

PROOF: First, suppose X is minimal curb but not tight. Then there exists some player $j \in$ N for whom $\beta_j(X) \subset X_j$ and $\beta_j(X) \neq X_j$. Let $X'_j = \beta_j(X)$ and $X'_i = X_i \forall_i \neq j$. Then $X' \subset X$, $X' \neq X$, $\beta_j(X') = X'_j$ and $M(X') \subset M(X)$, so $\beta_i(X') \subset \beta_i(X) \subset X'_i = X'_i \forall i \neq j$. Hence, $\beta(X') \subset X'$. The pay-off function u_j being continuous, the correspondence β_j from $M(X)$ to S_j is non-empty—and compact—valued, and, by Berge's Maximum Theorem, upper hemi-continuous.[3] Moreover, $M(X)$ is compact, so its image $\beta_j(X) = \beta_j(X') = X'_j$ is non-empty and compact, that is, $X' \in P$. In sum: X' is curb, contradicting the hypothesis that X is minimal. Thus, any minimal curb set is tight, and, being minimal among curb sets, it is minimal among tight curb sets. Secondly, suppose X is a minimal tight curb set. Applying the proof of Proposition 5.1 to the curb set X in the role of S, one establishes the existence of a minimal curb set $X' \subset X$. By the first part of the present proof, such a set X' is tight, and, since by hypothesis X is a minimal tight curb set, $X' = X$. Hence, X is a minimal curb set.

In order to relate the concept of a minimal curb set to established solution concepts, two further observations are useful, both being valid for all games G in $\mathcal{G}$. First, every curb set in such a game contains the support of at least one Nash equilibrium in mixed strategies. To see this, suppose $X \subset S$ is a curb set in $G = (N, S, U)$ and consider the game $G' = (N, X, U)$ obtained when players are restricted to the (non-empty and compact) strategy subsets X_i. Like G, the 'subgame' G' meets the conditions of the Glicksberg Theorem concerning the existence of Nash equilibrium in mixed strategies. Now, if $m \in M(X)$ is such an equilibrium of G', then it is also a Nash equilibrium of G, since by hypothesis each restricted strategy set X_i contains all best replies in S_i to strategies in X. Secondly, the set $R \subset S$ of rationalizable strategy profiles is non-empty and is the largest product set $X \subset S$ satisfying the equality $X = \beta(X)$. Hence, if the set R is compact, then $R \in P$ and R is the maximal tight curb set. Indeed, one can prove that R is compact (and non-empty) in games $G \in \mathcal{G}$ (Basu and Weibull 1990). In the light of this observation and Proposition 5.2, it should be clear why a minimal curb set and the set of rationalizable strategy profiles can be thought of as the two ends of a spectrum.

To illustrate these general findings, let us briefly return to Figure 5.1. One notes that the only curb set in 5.1 (a) is the full strategy space $\{T, M, B\} \times \{L, R\}$ *itself and it contains as a proper subset the support* $(\{T, M\} \times \{L, R\}$ itself, and it contains as a proper subset the support $(\{T, M\} \times \{L, R\})$ of the unique (mixed strategy) Nash equilibrium, which also happens to be quasi-strict. The only minimal curb set in 5.1b is $\{T\} \times \{L\}$, the support of the unique strict equilibrium. Note that the support of the undominated and hence perfect Nash equilibrium (M, R) is not contained in any minimal curb set.

Figure 5.2(a) shows the 'battle of the sexes' with three curb sets, $\{T\} \times \{L\}$, $\{B\} \times \{R\}$ and $\{T, B\} \times \{L, R\}$, all of which are tight, but only two of which are minimal. The non-minimal curb set appears to be a plausible pre-play 'agreement', viz., if the two players cannot agree on any of the two strict equilibria. The game in Figure 5.2(b) has two minimal curb sets, $\{T\} \times \{L\}$ and $\{M, B\} \times \{M, R\}$, the first containing a strict equilibrium and the second a non-strict equilibrium. Intuitively, the second set seems the more likely pre-play agreement.

	L	R
T	3,1	0,0
B	0,0	1,3

(a)

	L	M	R
T	1,1	0,0	0,0
M	0,0	2,3	3,2
B	0,0	3,2	2,3

(b)

	L	R
T	1,1	1,0
B	1,0	0,0

(c)

Figure 5.2

The game in 5.2(c) suggests that the curb requirement may in some games be too restrictive. Consider the set $\{T\} \times \{L\}$ in that game. This set is evidently not curb, since player 1 can costlessly deviate from T. Yet one could argue that it is 'closed' if not under 'rational' play, at least under 'rational and cautious' play, since the only other best reply for player 1 to 2's strategy L is his weakly dominated strategy B.

5.4. Sets Closed Under Rational and Cautious Behaviour

The above observation about optimal but weakly dominated strategies suggests the following weakening of the curb criterion. For each player $i \in N$, let $S_i^* \subset S_i$ be his subset of strategies that are not weakly dominated, that is, for which there exists no mixed strategy $m_i \in M(S_i)$ which weakly dominates s_i.[4] For any belief $m \in M(S)$, let $\beta_i^*(m) = \beta_i(m) \cap S_i^*$, the i'th player's undominated optimal strategies under m, and for any $X \in P$, let $\beta_i^*(X) = \beta_i(X) \cap S_i^*$. A set $X \subset S$ will be called closed under rational and cautious behaviour (curb*) if $X \in P$ and $\beta^*(X) \subset X$. In words: A set X of (pure) strategies is curb* if it is a non-empty and compact product set, and if the belief that strategies outside X will not be played implies that such strategies will indeed not be played by any player who is (a) rational in the sense of never playing strategies that are sub-optimal, and (b) cautious in the sense of never playing a weakly dominated strategy.[5] This weaker criterion is evidently met by the set $X = \{T\} \times \{L\}$ in Figure 5.2(c).

It is not difficult to verify that minimal *curb** sets exist in all games with continuous pay-off functions and compact strategy sets. The proof of Proposition, 5.2 does not apply, though. For while (in games $G \in \mathcal{G}$) the set $\beta(X)$ is non-empty and compact if X is, the set $\beta^*(X)$ need not be compact. However, one can show that it is non-empty for every $X \in P$.[6] Hence, if we use the slightly weaker tightness condition that X be contained in the closure of $\beta^*(X)$—such an X may be called almost tight—then a minor elaboration of the proof of Proposition 5.2 applies to *curb**, mutatis mutandis, leading to the following parallel conclusion: a set is a minimal *curb** set iff it is a minimal almost tight *curb** set.[7]

Notes

1. More exactly we define $u_i: S_i \times M(X) \to \mathbb{R}$ by $u_i(s_i \mid m) = \int U_i(z \setminus s_i)\, dm(z)$. Note that $u_i(s_i \mid m)$ is functionally independent of the component $m_i \in M(S_i)$.
2. In the language of Pearce (1984), a set $X \in P$ has the 'best-response property' if $X \subset \beta(X)$, so a curb set X is tight iff it has the best-response property.
3. We endow $M(S)$ with the *topology of weak convergence*. A sequence (m^t) from $M(S)$ is said to *converge weakly* to $m \in M(s)$ if $\int f d\, m^i \to \int f dm$ for all continuous functions $f: S \to \mathbb{R}$. Since S is a compact metric space, so is $M(S)$, see, for example, Theorem 6.4 in Parthasarathy (1967).
4. More exactly, but with a slight abuse of notation: there exists no $m_i \in M(S_i)$ such that $U_i(s' \setminus m_i) \geq U_i(s' \setminus s_i)$ for all $s' \in S$, with strict inequality for at least one $s' \in S$.

5. As argued in Weibull (1990), play of weakly dominated strategies can actually be both 'rational' and 'cautious' in games in which some player moves more than once in some play of the game. Hence, we take the present criterion to be generally valid only for games in which each player moves at most once in every play of the game.
6. For a proof of this claim, and for an example of a non-compact set $\beta^*(X)$, see Basu and Weibull (1990).
7. More precisely, we define a set $X \subset S$ to be an almost tight curb* set if $X \in P$ and $X = \beta^*(X)$. Note that if $X \in P$ and $\beta^*(X) \subset X$, then $\beta^*(X) \subset X$.

REFERENCES

Basu, K. and J.W. Weibull. 1990. 'Strategy Subsets Closed under Rational Behaviour'. Princeton University. Princeton, NJ. Discussion paper no. 62.

Bernheim, D. 1984. 'Rationalizable Strategic Behaviour'. *Econometrica* 52: 1007–29.

Kohlberg, E. and J.F. Mertens. 1986. 'On the strategic stability of equilibria'. *Econometrica* 54: 1003–37.

Parthasarathy, K. 1967. *Probability Measures on Metric Spaces*. New York: Academic Press.

Pearce, D. 1984. 'Rationalizable Strategic Behaviour and the Problem of Perfection'. *Econometrica* 52: 1029–50.

Weibull, J.W. 1990. 'Self-enforcement in Extensive-form Games'. Institute for International Economic Studies, Stockholm University, Stockholm Seminar paper no. 470.

6 Group Rationality, Utilitarianism, and Escher's Waterfall

6.1. Motivation

There is a long tradition in linguistic and moral philosophy that claims that language has at least two uses. First, there is the 'normal' or 'serious' use which allows us to communicate and exchange information. For its normal use what matters is 'the meaning of what is said'. Habermas (1989: 159) has described this as 'the original mode of language'. Secondly, there is what has been described as the 'etiolated' or 'parasitical' use (Austin 1962) whereby language is used by the speaker to *bring about* something in the world. I shall here make a related distinction, between normal and 'honest' speech, on the one hand, and parasitical and 'strategic' speech on the other. If a murderer asks me about the whereabouts of a potential victim and I say, 'He has gone left', knowing full well that he has gone right, I am using language strategically in order to bring about a certain kind of world. Note that if language were never used normally or honestly, then my saying 'He has gone left' would have no effect.

It seems possible, therefore, to argue that if we made excessive use of language in a parasitical way for strategic purposes, it would lose much of its value even as a normal instrument of communication. Now consider a world in which everyone is committed to the same objective function. While the objective function could be of any kind, let us for the sake of an example suppose that all individuals are committed utilitarians; and they practise this to the point where before uttering each sentence they calculate the total utility that

From *Games and Economic Behaviour,* 7, 1994: 1–9.

I am grateful to Abhijit Banerjee, John Conway, Uli Hege, David Lewis, Robert Pollak, Andrew Postlewaite, Ariel Rubinstein, Ekkehart Schlicht, Siddharth Sahi, and, especially, Debraj Ray for comments and discussion. I have also benefited from seminars at Princeton, Cornell, the Institute for International Economic Studies at Stockholm, and the University of Pennsylvania.

will be generated as a consequence of it and then make those statements which generate the most utility. It seems possible to argue that such fastidious adherence to utilitarianism could be harmful for society in terms of utilitarian calculations itself, because, in such a world, language would cease to have power, communication would break down and so would many things which are predicated upon our ability to communicate. Such an argument, where utilitarianism is pitted against utilitarianism, has been made in the literature. Hodgson (1967) is quite explicit on this.

But is it possible to build a formal argument where utilitarianism gets pitted against utilitarianism? The purpose of this chapter is to answer this question. It is shown that if the number of decisions to be taken in a society is finite then the answer must be negative. In an infinite-decision society it is possible that each person acting in utilitarian interest everywhere will end up creating a society inferior in utilitarian terms itself.

Though I began with the case of language and utilitarianism, the point that this chapter tries to illustrate is a more general one. Consider a game in which each player can choose from two possible strategies, S and H, and all players have the same pay-off function. This could be because all players belong to a team with one common objective or because they are moral creatures, all committed to one social welfare function.[1] The question being asked here is whether this game can have a dominant-strategy equilibrium which is Pareto inferior. More specifically, is it possible that whenever an individual switches from playing H to S (other players' choices remaining unchanged) he, and therefore everybody, becomes better-off but if everybody played S everybody would get a lower pay-off than what they would get if everybody played H? The answer to this is yes and this is demonstrated in Section 6.2 by constructing a game with a 'paradoxical' outcome, which will be referred to here as the 'Waterfall' paradox. A more difficult problem is to demonstrate this possibility for games which treat players 'symmetrically', in a sense made precise below. This is demonstrated in Section 6.2 under the assumption that the axiom of choice is true. Section 6.3 discusses some implications of these 'Waterfall' games for models of economics.

6.2. 'Waterfall' Games

The result that will be established first is this. If we have an infinite number of players then there exists a game in which everybody has the same objective function (that is, they may, for instance, be maximizing player 1's utility or all their utilities summed up) but each individual's effort to maximize this objective function ends up with the players in a sub-optimal equilibrium, sub-optimality being defined in terms of the same objective function.

Let me make this more precise. Suppose N is the set of players or individuals, and Ω is the power set of N. Each player $i \in N$ can play one of two available strategies: H and S. Player i's pay-off function is f_i. That is, $f_i \ \Omega \to \mathrm{R}$, where R is the set of real numbers. This is to be interpreted as follows. If $A \in \Omega$

is the set of players who choose S (i.e., $N \setminus A$ is the set choosing H), then i's pay-off is $f_i(A)$. Since everything else in the description of a game is constant, in this chapter we identify a game with a specification of pay-off functions $\{f_i\}_{i \in N}$.

We shall describe a game, $\{f_i\}_{i \in N}$, as a prisoner's dilemma if each player prefers to play S no matter what others play (i.e., S is a dominant strategy for each player), but everybody is worse off if everybody plays S as compared to the case where everybody plays H.

That prisoner's dilemma games exist for N finite or infinite is well-known and can be proved by construction. The question with which this chapter is concerned is, however, more difficult to answer.

Suppose all players have the same pay-off function. That is, there exists a real-valued function, f, on Ω such that, for all $i \in N$, $f_i = f$. I shall refer to such a game as a *uniform game*. This could happen if all of them are perfectly united in their objective. For instance, they may be card-carrying utilitarians with each person acting so as to increase the sum of their happiness or they could be a group of firms in an industry, totally united in their pursuit of maximizing joint profits or they may be employees of a firm all striving to maximize the firm's profit, as assumed in most economics textbooks. The first question I want to address is this. Does there exist a uniform game which is a prisoner's dilemma?

The answer is yes. I shall construct an example in a moment to demonstrate this. The proof requires that the number of players be infinite. A much more difficult question is: whether there exists a game which is uniform, a prisoner's dilemma, and symmetric in pay-offs (in a sense to be made precise later)? Even here the answer is 'yes' but my proof of this will make use of the axiom of choice.

Given that in the social sciences we often resort to the assumption of an infinite number of agents,[2] these games are not only of intrinsic interest but of relevance to our models.

Note first that with a finite number of agents, no matter how large, there cannot be a game which is uniform and a prisoner's dilemma. To see this consider two elements, A and B in Ω, such that outcome A implies everyone plays H and B implies everyone plays S: that is, $A = \phi$, $B = N$.

Since there is a finite number of players we can go from A to B by changing H to S for one player at a time. If H is a dominated strategy for each player, every such switch must raise the pay-off (remember in a uniform game all players have the same pay-off function). Hence, $f(B) > f(A)$. So the game cannot be a prisoner's dilemma.

Now it will be shown that if N were infinite, the above paragraph's claim would not be true. Somewhat surprisingly, this is so even if N is countably infinite. In fact, from here onward, I assume $N = \{1,2,...\}$; that is, N is the set of all positive integers. In other words it is possible that, beginning with any outcome, as each H is changed to an S, the pay-off rises, but the pay-off from everyone playing S is less than the pay-off from the case where everyone plays H. A picturesque analogue of this is Escher's well-known painting, 'Waterfall', in which at each step water keeps flowing down but ends up on top.

Since in a uniform game all players have the same pay-off function, such a game is identified entirely by the pay-off function, $f: \Omega \rightarrow R$. Hence the main proposition being proved is this:

There exists a uniform game, f, such that

(6.1) For all $A \in \Omega$, and for all $i \in N \backslash A$, $f(A \cup \{i\}) > f(A)$ and

(6.2) $f(N) < f(\phi)$.

This is proved by actually constructing an example. Let $\hat{f}$ be the following pay-off function. If $A \in \Omega$, $\hat{f}(A)$ is defined as follows. If $\# N \backslash A < \infty$ (i.e., A is such that the number of players who choose H is finite) then $\hat{f}(A) = -\# N \backslash A$. Next suppose $\# N \backslash A = \infty$. Then $\hat{f}(A)$ is the number (in decimals) formed by writing 1 in the ith place after the decimal if $i \notin A$ (i.e., player i has chosen H), and 2 in the ith place after the decimal if $i \in A$. To avoid misunderstanding, let me state this a little more formally. Given $A \in \Omega$, define

$$x_i(A) = \begin{cases} 2 & \text{if } i \in A \\ 1 & \text{if } i \notin A. \end{cases}$$

If $\# N \backslash A = \infty$, then $\hat{f}(A) = \cdot x_1(A) x_2(A) \ldots$, where $\cdot x_1(A) x_2(A) \ldots$ is a decimal number where the number in the ith place after the decimal is $x_i(A)$.

It follows that, $\hat{f}(N) = 0$ and $\hat{f}(\phi) = .111\ldots$. Hence, $\hat{f}$ satisfies (6.2). It is easy to check that it satisfies (6.1) as well.

Observe that in the game, $\hat{f}$, the impact on the pay-off of a player's decision between H and S depends on who the player is. Player 1, for instance, has a much bigger effect on the pay-off, than player 58. Now, let us suppose that we want the game to be symmetric or anonymous. Some of the applications I discuss in the next section make anonymity a reasonable property. But a pay-off function can be anonymous in several senses. I shall here consider a weak anonymity property, which will be called finite-anonymity.

Earlier a particular 'play' of the game was denoted by a set $A \in \Omega$. There is an alternative characterization according to which a play is a sequence $\{x_i\}_{i \in N} \equiv \{x_i\}$, in which $x_i \in \{H, S\}$, for all i. The jth element in $\{x_i\}$ denotes what player j has chosen. Clearly we could think of a function, h, which for every play, $\{x_j\}$, specifies the set of players who have chosen S. That is $h(\{x_j\}) = \{i \in N \mid x_i = S\}$. The pay-off from $\{x_i\}$ will be, distorting earlier terminology a little, written as $f(\{x_i\})$. If we were more cautious we would write $f(h(\{x_i\}))$ instead of $f(\{x_i\})$.

Next, we need to define a finite permutation. The mapping $\sigma: N \rightarrow N$ is a *finite permutation* if there exists a finite set $A \subset N$ such that, for all $i \notin A$, $\sigma(i) = i$ and the restriction of σ on A is a permutation (i.e., it is one-to-one and onto on A).

Finally, we say that the uniform game f satisfies *finite anonymity* if the following is true: If $\{x_i\}$ and $\{y_i\}$ are plays that there exists a finite permutation, σ, such that $\{x_i\} = \{y_{\sigma(i)}\}$, then $f(\{x_i\}) = f(\{y_i\})$.

The second, and main, paradoxical result is this: There exists a uniform game which satisfies (6.1), (6.2), and finite anonymity.

Interestingly, it does not seem possible to give a constructive proof of this. In other words, what is being claimed is that though I cannot construct such a game, I can prove that such a game must exist.

In order to prove this define the binary relation, $\sim$, on Ω as follows: For all $X, Y \in \Omega$, $X \sim Y$ iff $\# X \backslash Y$ and $\# Y \backslash X$ are both finite. It will now be shown that $\sim$ is an equivalence relation. Its reflexivity and symmetry are obvious. To check that $\sim$ is transitive, suppose $X \sim Y$ and $Y \sim Z$. Hence $\# X \backslash Y$, $\# Y \backslash X$, $\# Y \backslash Z$, and $\# Z \backslash Y$ are finite. It is easy to see that $X \backslash Z \subset (X \backslash Y) \cup (Y \backslash Z)$. Let $x \in X \backslash Z$. If $x \notin Y$, then $x \in X \backslash Y$. Suppose $x \in Y$. Since $x \notin Z$, it follows $x \in Y \backslash Z$.

Hence $\# X \backslash Z$ is finite and by an anlogous proof $\# Z \backslash X$ is finite.

Since $\sim$ is an equivalence relation, there exists a partition P of Ω generated by $\sim$ such that if there exists $P_\alpha \in P$, such that $X, Y \in P_\alpha$ then $X \sim Y$.

Let Q be a set formed by choosing one element from each element of P. The existence of this set is guaranteed by the axiom of choice.

Define $g : Q \rightarrow R$ as follows:

Let $A \in Q$. Hence there exists a unique element, P_α, in P such that $A \in P_\alpha$.

(a) If $\phi \notin P_\alpha$, and $N \notin P_\alpha$, then $g(A) = 1$.
(b) If $\phi \in P_\alpha$, then $g(A) = \# A + 1$.
(c) If $N \in P_\alpha$, then $g(A) = - \# (N \backslash A)$.

Note that ϕ and N cannot belong to the same element of the partition. They belong to the collection of, respectively the finite sets and the cofinite sets. Also, we could have chosen ϕ and N from these sets in constructing Q. In what follows it is worth keeping in mind that A and $A \cup \{i\}$ are always in the same element of the partition.

Now let $f^*: \Omega \rightarrow R$ be an extension of g defined as follows. Let $A \in \Omega$. Find a $B \in$ Q such that $A, B \in P_\alpha \in P$, for some P_α. We set $f^*(A) = g(B) - \# (B \backslash A) + \# (A \backslash B)$.

It is easy to check that f^* satisfies (6.1), (6.2), and finite anonymity. I shall here demonstrate finite- anonymity, since (6.1) and (6.2) are obvious.

Let $x \equiv \{x_i\}$ and $y \equiv \{y_i\}$ be such that there exists a finite permutation σ of N such that $\{x_i\} = \{y_{\sigma(i)}\}$. Hence it follows that $\# X \backslash Y = \# Y \backslash X < \infty$, where $X \equiv \phi\,(x)$ and $Y \equiv \phi\,(y)$. Therefore, $X, Y \in P_\alpha \in P$, for some P_α. Let $B \in P_A \cap Q$. It is easy to see that $\# X \backslash B - \# B \backslash X = \# Y \backslash B - \# B \backslash Y$. This follows from the fact that

$$\# X \backslash B + \# B \backslash Y = \# X \backslash Y + \# (X \cap Y) \backslash B + \# B \backslash (X \cup Y), \text{ and}$$

$$\# Y \backslash B + \# B \backslash X = \# Y \backslash X + \# (X \cup Y) \backslash B + \# B \backslash (X \cup Y).$$

Hence, $f^*(X) = f^*(Y)$, thereby establishing that f^* satisfies finite anonymity.

It may be asked as to why attention is restricted to *finite* anonymity. This is because it is easy to show that a *fully* anonymous uniform game is incompatible with (6.1), let alone (6.1) *and* (6.2). Define $r : N \rightarrow N$ to be a *permutation* if r is one-to-one and onto on N. The uniform game f is *fully anonymous* if whenever $\{x_i\}$ and $\{y_i\}$ are plays such that for some permutation r, $\{x_i\} = \{y_{r(i)}\}$, then $f(\{x_i\}) =$

$f(\{y_i\})$. Now let $\{x_i\}$ and $\{y_i\}$ be such that $x_i = H$ if and only if i is odd and $y_i = x_i$, if $i > 1$, and $y_1 = S$. By (6.1), $f(\{y_i\}) > f(\{x_i\})$. But note that there exists a permutation r such that $\{x_i\} = \{y_{r(i)}\}$. This is so if

$$r(i) = \begin{cases} i + 2, & \text{if } i \text{ is odd} \\ i - 2, & \text{if } i \text{ is even but not } 2 \\ 1, & \text{if } i = 2 \end{cases}$$

Hence, $f(\{x_i\}) = f(\{y_i\})$, which is a contradiction.

It would be interesting to develop notions of anonymity which lie between the polar extremes of finite and full anonymity and examine their compatibility with (6.1) and (6.2).

6.3. Remarks

What the 'Waterfall' game demonstrates is, in some sense, the opposite of the widsom embodied in the well-known invisible-hand theory. According to the latter, every individual working in his individual interest may lead to an outcome which is optimal for the group. According to the 'Waterfall' game, every individual working in the group's interest may lead to an outcome which is sub-optimal for the group. For models of economics the results of Section 6.2 are important because such models do often assume an infinity of agents. One response to this is to treat the paradoxical 'Waterfall' effect as a *reductio ad absurdum* against the idea that, if a population is large and individuals are insignificant, we can safely model the population as infinite.[3] This would amount to a critique of models like that of perfect competition in economics.

There are three remarks in this connection worth keeping in mind. First, for certain kinds of issues, an infinity of agents may not be an unrealistic assumption, as long as it is countable. This is because if we consider all people of present and future generations the number may well be infinite.

Secondly, in anonymous games the 'Waterfall' effect seems to be a consequence of not just an infinite number of agents but also the axiom of choice. Hence, it may be possible to avert the paradoxical result by foregoing the axiom of choice, and indeed there are some areas of game theory which have tried to do without the axiom.

Finally, there is an interpretation under which the results of Section 6.2 could be thought of as occurring in models with a finite number of agents. Suppose there are n (finite) agents and each agent i takes t_i decisions. Each decision consists of choosing between S and H. For instance, an agent i might be having t_i points of time between now and next year at each of which he has to decide whether to do something (H) or procrastinate (S) to the next point of time, like in Akerlof's (1991) model. As long as $t_i + \ldots + t_n$ is infinite, the same construction as in Section 6.2 is possible. It is important, however, to note that in this case there must exist one agent who controls an infinite number of decisions. That is, for some i, t_i must be infinite. Hence, in this case S being chosen everywhere is not a dominant strategy for player i. However, S being chosen everywhere could be thought of as *decision-wise dominant*, that is, for *each* of the t_i decisions, considered one at a time, S dominates H.

To close with a digression, I turn to a problem in welfare economics on which the 'Waterfall' game throws some light. The distinction between act and rule consequentialism has always been considered ambiguous[4] and I have shared in the feeling. *Act* consequcntialism requires that a particular act or decision should be undertaken if it brings about a social state which is desirable. *Rule* consequetialism recommends an act, if it is the case that everyone undertaking the act in similar situations leads to a desirable social state. The distinction between the two, however, seems questionable because by making the definition of what constitutes a 'similar situation' sufficiently specific, we can make rule consequentialism operationally indistinguishable from act consequentialism.[5]

Our exercise in Section 6.2 clarifies the conditions under which the two moral systems can be distinguished. If the uniform game is given by the f^* defined above, then act utilitarianism would recommend that individuals use speech strategically (that is, choose S), whereas rule utilitarianism would clearly not make such a recommendation. If a 'rule' is taken to mean either of the following two recommendations: (i) Be strategic (S) and (ii) be honest (H), then rule utilitarianism would recommend honesty. But if we allow rules to be more complicated and take forms like 'in the following circumstances, be honest' then rule utilitarianism may make more complicated recommendations. The argument of Section 6.2 may be seen as pushing us away from act consequentialism of any form, including utilitarianism, towards some form of deontological ethics because it highlights how act consequentialism can be self-defeating. The argument also shows that in some sense rule consequentialism is closer to deontological ethics than act consequentialism.

NOTES

1. It is, therefore, immediately obvious that my reference to utilitarianism above is an illustration of any consequentialist ethical system. It is, however, worth noting that some have considered the distinction between consequentialism, in general, and utilitarianism, in particular, to be blurred. Hammond (1990: 2), for instance, argues that 'utilitarianism itself can be derived from the even more primitive principle called "consequentialism" '.
2. In industrial organization theory the model of perfect competition and in general-equilibrium theory Walrasian analysis are often rationalized in terms of an infinite number of firms and agents (Aumann 1964; Hildenbrand and Kirman 1976). In fact, while these models require a continuum of agents, the paradoxical result described here occurs even with a countably infinite set of agents. I write 'even' because, as will be obvious later, my result is easily generalized to the uncountable case, though I work with the countable one.
3. This is the position which David Lewis seems to take (personal communication to the author, dated 15 January 1990).

4. See Smart (1973). His discussion is concerning act and rule utilitarianism but the same ideas are easily extended to the more general concept of consequentialism (see Sen 1985).
5. See Smart (1973: 9–12) for discussion.

REFERENCES

Akerlof, G. 1991. 'Procrastination and Obedience'. *American Economic Review* 81 (Papers and Proceedings).

Aumann, R.J. 1964. 'Markets with a Continuum of Traders'. *Econometrica* 32.

Austin, J.L. 1962. *How to do Things with words*. London/New York: Oxford University Press.

Habermas, J. 1989. *On Society and Politics: A Reader*. Boston: Beacon Press.

Hammond, P.J. 1990. 'Interpersonal Comparisons of Utility: Why and How They are and Should be Made'. University of Florence. Mimeo.

Hildenbrand, W. and A.P. Kirman 1976. *Introduction to Equilibrium Analysis*. Amsterdam: North-Holland.

Hodgson, D.H. 1967. *Consequences of Utilitarianism*. London/New York: Oxford University Press.

Sen, A. 1985. 'Well-Being, Agency and Freedom'. *Journal of Philosophy* 82.

Smart, J.J.C. 1973. 'An Outline of a System of Utilitarian Ethics'. In J.C.C. Smart and B. Williams, eds. *Utilitarianism: For and Against*. London/New York: Cambridge University Press.

7 Information and Strategy in Iterated Prisoner's Dilemma

The prisoner's dilemma (PD)—a two-person, non-zero-sum game—highlights a situation where individual rationality leads to a collectively sub-optimal equilibrium. This has disturbing implications for the social sciences; and the problem has received considerable attention from philosophers, psychologists, and economists. It has been used to interpret Rousseau's concept of the 'general will' and has also cropped up in discussions on ethics and morality.[1] It demonstrates with disconcerting simplicity how atomistic action may lead to collective ill, unless there are binding 'social contracts' to guide individual action.

Respite was sought by pointing out that in reality this game was played more than once; and that could result in the equilibrium not being a socially inferior one. But it was supposedly demonstrated by Luce and Raiffa (LR) (1958) that if the game was repeated a finite number of times, then the equilibrium strategy continued to be the sub-optimal one. It will be argued here that this is not in general so and for their result to hold, it is necessary for certain additional assumptions—which are not required in the non-repeated prisoner's dilemma—to be satisfied. These assumptions are highly demanding and proliferate at a remarkable pace as the number of repetitions is increased. This is an optimistic result. It shows that in reality, where this game situation is likely to arise more than once, individual rationality may lead to collective good.

7.1

The following table is the pay-off matrix for individuals A and B. a_1, a_2 and b_1, b_2 are possible strategies for A and B, respectively, and α_{ij} (respectively, β_{ij}) is

From *Theory and Decision* 8, 1977: 293–8.
I have benefited from discussions with Professor Amartya Sen.

the pay-off to A (B) for (a_i, b_j). For simplicity we consider the pay-offs to be monetary payments.

		B	
		b_1	b_2
A	a_1	α_{11}, β_{11}	α_{12}, β_{12}
	a_2	α_{21}, β_{21}	α_{22}, β_{22}

Consider the following assumptions:

(i) $\alpha_{21} > \alpha_{11} > \alpha_{22} > \alpha_{12}$,

(ii) $\beta_{12} > \beta_{11} > \beta_{22} > \beta_{21}$.

Individual A is said to have prisoner's dilemma preference if (i) holds and he maximizes monetary gains. Similarly for B. We shall assume throughout that both A and B have prisoner's dilemma preferences. Consequently, if this game is played once only, then (a_2, b_2) is the equilibrium outcome. This is Pareto inferior to (a_1, b_1). Therefore, if this sort of a preference configuration occurs in reality, as seems quite plausible, then a Pareto inefficient outcome seems inevitable.

However, in reality most games are played more than once. Suppose that the individuals know that the above game will be played more than once. Then A might play a_1 a few times to induce B to play b_1. Once B gets the hint and plays b_1, the equilibrium might settle at (a_1, b_1). A (resp. B) would not break this equilibrium, since though by suddenly playing a_2 (resp. b_2) he can reap some extra gains, B (resp. A) will retaliate with b_2 (resp. a_2) making A worse off eventually.[2] This seems to provide a way out of the Pareto inferior outcome.

However, if the game is played a finite number of times, and the players know how many times, then this 'way out' is allegedly illusory. This paradoxical result was demonstrated by Luce and Raiffa (1958: 98–9) using an argument similar to that used in the 'surprise-test' paradox:

Suppose the players are told that game H (i.e. the above game) is to be played exactly, twice, and suppose that each player is shrewd enough to see that his second strategy, strictly dominates his first one in a single play of the game. Thus, before making their first move, each realizes that in the second game the result is bound to be (a_2, b_2), for, after the first game is played, the second one must be treated as if H is going to be played once and only once. The second play being perfectly determined, the first play of the game can be construed as H being played once and only once. Thus, it appears that (a_2, b_2) must arise on both trials. The argument generalizes: Suppose they know that H is to be played exactly 100 times. Things are clear on the last trial, the (a_2, b_2) response is assured; hence the penultimate trial, the 99th, is now in strategic reality the last, so it evokes (a_2, b_2); hence the 98th is in strategic reality the last, so it evokes (a_2, b_2) etc. This argument leads to (a_2,

b_2) on all hundred trials. Indeed if player 2 is a b_2 -conformist on all trials, then player 1 is best off choosing a_2, on all trials, and conversely, i.e., (a_2, b_2) on all trials is an equilibrium pair.

From the above argument it seems that given the assumption that the two individuals have prisoner's dilemma preferences, the equilibrium is (a_2, b_2) whether the game is played once or any finite number of times. I would argue that this is incorrect. The mere assumption that *A* and *B* have prisoner's dilemma preferences is sufficient to establish (a_2, b_2) as the equilibrium in the single game case. But when the game is repeated N (> 1) times, the assumption requirement is much stronger. This can be seen by carefully analysing the case of N games.

Consider the case of $N = 2$. Luce and Raiffa argued that in the first game *A* will move a_2 because *A* knows that in the final game (a_2, b_2) is bound to occur. But how does *A* know that? Clearly, to know that, *A* must know that *B* has prisoner's dilemma preference. Symmetrically, *B* must know that *A* has prisoner's dilemma preference. Therefore, to prove (a_2, b_2) to be the equilibrium in the first game we need this additional assumption that *A* (resp. *B*) knows that *B* (resp. *A*) has prisoner's dilemma preference.

Now, even if $N = 100$, this assumption is necessary since, without it, we would not be able to prove that (a_2, b_2) would certainly occur in the 99th game. Hence, for all $N > 1$, this assumption is necessary. But is this assumption sufficient for all N ? The answer is no. As N becomes larger, the list of assumptions increases.

Consider $N = 3$. We can be sure that *A* will move a_2 in the first game if he thinks that outcomes of games 2 and 3 are fixed. *A* will think in this way if (1) *A* knows that *B* has prisoner's dilemma preference (this will make *A* sure about the final game's outcome to be (a_2, b_2) and (2) *A* knows that *B* knows that *A* has prisoner's dilemma preference. This will make *A* sure that *B* will move b_2 in the second game, and consequently will make him sure that the outcome of the second game will be (a_2, b_2). Remember that because of (1), *A* will move a_2 in the second game. Given that symmetric conditions hold for *B*, the outcome of the first game will be (a_2, b_2).

The increasing informational assumption is obvious. We now summarize the above analysis.

When $N = 1$, (a_2, b_2) is the equilibrium if,

A and *B* have prisoner's dilemma preferences.

When $N = 2$, (a_2, b_2) would certainly occur in both games if *A* and *B* have prisoner's dilemma preferences, and

A (resp. *B*) knows that *B* (resp. *A*) has prisoner's dilemma preference.

When $N = 3$, (a_2, b_2) would certainly occur in all three games if

A and *B* have prisoner's dilemma preferences,

A (resp. *B*) knows that *B* (resp. *A*) has prisoner's dilemma preference, and

A (resp. *B*) knows that *B* (resp. *A*) knows that *A* (resp. *B*) has prisoner's dilemma preference.

What happens in the general case where $N = n$? The result is summarized in the Table 7.1. It is assumed throughout that both A and B have prisoner's dilemma preferences.

Table 7.1

	A *moves* a_2, *if the following additional assumptions are given*	B *moves* b_2, *if the following additional assumptions are given*
*n*th time	No additional assumptions	No additional assumptions
(n–1)th time	(1) A knows B has prisoner's dilemma preference	(1*) B knows A has prisoner's dilemma preference
(n–2)th time	(1) and (2) A knows that(1*)	(1*) and (2*) B knows that (1)
(n–3)th time	(1), (2) and (3) A knows that (1*) and (2*)	(1*), (2*) and (3*)B knows that (1) and (2)
.		
.		
.		
(n-(n–l)th time (i.e. 1st time)	(1), (2) ... ((n–2)) and ((n–1)) (A) knows that (1*), (2*) ... and ((n–2)*)	(1*), (2*)((n–2)*) and ((n–l)*) B knows that (1), (2) ... and ((n–2))

Hence, while in the case of non-repeated prisoner's dilemma the sub-optimal solution is a direct consequence of the individuals' having prisoner's dilemma preferences, in the n-game case it is, fortunately, not so.

While people are often confronted with prisoner's dilemma situations in life, it is not at all clear that they can be expected to fulfil assumptions (1), (2) ... (n–l) and (1*), (2*) ... ((n–l)*). These are extremely strong informational requirements. When these assumptions are not satisfied, the occurrence of (a_1, b_1) in some games becomes a distinct possibility. Under what parametric conditions the 'general will' is likely to be realized is discussed in the case of $N = 2$ in the next section.

However, it is already clear that though a divergence in the 'general will' and the 'will of all' could create serious problems as discussed by Runciman and Sen (1965) the occurrence of such a divergence may not be as widespread as one may have been led to believe from the Luce and Raiffa (1958) argument.

7.2

We assume $N = 2$; A and B have prisoner's dilemma preferences and in situations of uncertainty they maximize their individual *expected* earnings; $\alpha_{ij} = \beta_{ji}$, for all i, j (i.e. A and B have symmetric pay-offs); and A's expectation of B's behaviour is the same as B's expectation of A's behaviour. These last two assumptions are not really necessary but are made to allow us to restrict our analysis to one individual, since the other will behave symmetrically. Finally, no assumptions like (1), (2), ... ((*n*–1)) and (1*), (2*), ... ((*n*–1)*) are granted.

Since *A* and *B* have prisoner's dilemma preferences, (a_2, b_2) is bound to occur in the final game. Hence, our main interest lies in the first game. Consider the subjective probabilities attached by *A* to *B*'s moves. The probability of *B* moving b_2 in the first game is *r*. If in game 1 *A* moves a_2, then the probability of *B* playing b_2 in the second game is *p*. If in game 1 *A* moves a_1, then the probability of *B* playing b_2 in the second game is *q*. If *A* moves a_2 in game 1, it is natural for him to expect retaliation from *B* in the second game. So we may assume $p > q$, though this assumption is not necessary for the subsequent results.

Let *A*'s expected gain from playing a_1 in the first game be *G*(1).
Hence,

$$G(1) = (1-r)\,\alpha_{11} + r\,\alpha_{12} + (1-q)\,\alpha_{21} + q\,\alpha_{22}.$$

Similarly, the expected gain from playing a_2 in the first game is *G*(2).

$$G(2) = (1-r)\,\alpha_{21} + r\,\alpha_{22} + (1-p)\,\alpha_{21} + p\,\alpha_{22}.$$

In both these expressions, the first two terms are expected earnings from game 1, and the last two terms are expected earnings from game 2. Because of our assumption above, *B*'s subjective probabilities regarding *A*'s behaviour are symmetric. Hence, a similar analysis applies to *B*.

Let $X = G(1) - G(2)$. Therefore,

$$X = (1-r)(\alpha_{11} - \alpha_{21}) + r(\alpha_{12} - \alpha_{22}) + (p-q)(\alpha_{21} - \alpha_{22}).$$

If $X > 0$, then *A* plays a_1 in game 1; and if $X < 0$, then *A* plays a_2. In case $X = 0$, then *A*'s move is indeterminate in the context of this theory. It is easily checked that suitable values for the parameters involved can be selected such that (i) is satisfied and at the same time $X > 0$. Hence, it is possible for *A* to have prisoner's dilemma preference and at the same time move a_1, in the first game. *B*'s behaviour being symmetric, the occurrence of (a_1, b_1) is a distinct possibility.

Assumption 1 (p. 296) says that $p = q = 1$. If this assumption holds, then $X < 0$; and in line with the analysis in the first section, *A* will move a_2 in both games. But the assumption is an arbitrary one

As *N* becomes larger, the assumption structure required to establish (a_2, b_2) as the equilibrium becomes extremely bulky and untenable in any realistic situation. Consequently, the possibility that individual rationality may lead to the 'common good', even when individuals have prisoner's dilemma preferences, is no longer as remote as it might have seemed.

NOTES

1. For the former see Runciman and Sen (1965); and Smyth (1972); and for the latter see Sen and Watkins (1974).

2. For a more detailed exposition of this see Luce and Raiffa (1958:98).

Reference

Luce, R.D. and H. Raiffa. 1958. *Games and Decisions*. New York: Wiley.

Runciman, W.G. and A.K. Sen. 1965. 'Games, Justice and the General Will'. *Mind* 74.

Smyth, J. 1972. 'The Prisoner's Dilemma II'. *Mind* 81.

Sen, A.K. 1974. 'Choice, Orderings and Morality'. In S. Korner, ed. *Practical Reason*. Oxford: Blackwell.

Watkins, J.W.M. 1974. 'Comment: "Self-Interest and Morality"'. In S. Korner, ed. *Practical Reason*. Oxford: Blackwell.

Part III
Industrial Organization and Strategic Behaviour

8 Monopoly, Quality Uncertainty, and 'Status' Goods

8.1. Introduction

To buy certain products where quality is important, for example Jaguar cars, the newly released shares of a company, meals at a popular Chinese restaurant, or consultation with a 'good' doctor, one has often to join a waiting list.[1] In other words, these 'firms' price their products so as to maintain an excess demand. At first sight this is paradoxical. Why does the doctor not raise his fees and eliminate or shorten the waiting line of patients? While this *may* be attributed to the benevolent spirit of the medical profession, the present chapter provides an explanation of such behaviour in terms of selfishness and, in particular, profit maximization. It presents a theory of why and under what circumstances it pays a firm to follow a strategy of maintaining excess demand.

There is a long tradition in economics—dating back to at least Scitovsky's paper of 1944—of modeling markets where consumers judge quality by price. In a situation of price rigidities, a natural analogue of this is that consumers will judge quality by also observing the aggregate excess demand for the good. At first sight, this seems to provide an explanation of why a firm may wish to maintain a price at which there is an excess demand: because by this it manages to increase the desirability of its product to its individual consumers.

This argument, however, turns out to be inadequate if demand functions are continuous. Consider a monopoly. In the presence of excess demand its sales need not be affected by small changes in price. In other words, it is possible for the firm to raise price a little without having to sell less, thereby increasing its profit. Hence, an excess-demand situation cannot be a profit-maximizing one.

From *International Journal of Industrial Organization* 5, 1987: 435–46.

This problem was suggested to me by Avinash Dixit. A first version of this chapter was written at the Institute for Advanced Study, Princeton.

Note that the above reasoning makes use of the continuity of demand. Hence, *if it were the case* that people judge quality by excess demand *and* the demand function in discontinuous, only then may it be possible to explain excess demand in equilibrium. What is interesting, and this is the central point of this chapter, is that if people actually judge quality by excess demand, then the demand function automatically turns out to be discontinuous and under certain conditions this discontinuity is precisely of the kind that results in price rigidities and excess demand. It is important to emphasize that the discontinuity of demand which plays a crucial role in this chapter is *not* an assumption but arises naturally in this framework. It arises even though the utility function or the *n*-function (defined below), i.e., all the basic ingredients of the model, are continuous.

While the model in this chapter is motivated by considering products where quality is judged by excess demand, there are several other situations where the formal model would be applicable. First, it can be applied to the supply side. For instance, by thinking of jobs as work environments or certificates of prestige which people buy, my argument can be inverted to explain wage rigidities in the face of excess supply of labour. Secondly, it could apply to markets where the quality of the product is not in question but where people buy a good or service to establish their own social status or worth. The membership of a club could often be a symbol of status. Calcutta's old colonial Calcutta Club provides an example of this. One reason why an individual covets the membership of this club is because there are others who covet the same but nevertheless fail to become members. Aggregate excess demand for membership enhances an individual's value of the membership. These alternative motivations for the same formal model are discussed in Section 8.4. The next two sections develop the model and its properties. In the present chapter the problem is analysed within one specific industrial structure, namely, a simple monopoly.

8.2. The Model

Let $H = \{1, \ldots, h\}$ be the (finite) set of individuals. It will be assumed that there is only one producer of a certain good, for example, cars. Each person i in H can buy at most one car, and the maximum amount of money, v_i, that he is willing to pay for the car depends on the excess demand (i.e., the difference between aggregate demand and aggregate supply), z, that exists: hence, for all i in H,

$$(8.1) \qquad V_i = v_i(z),\ v_i' \geqq 0.$$

If p is the price of a car and z the excess demand for it, a person i will want to buy the car if and only if $v_i(z) \geqq p$, and the total demand for cars, n (which is equal to the total number of people who want to buy), is given by[2]

$$(8.2) \qquad n = n(z, p) = \#\{i \in H \mid v_i(z) \geqq p\}.$$

Given (8.1), it follows that as z increases or p falls, n will not fall. Since each individual can buy at most one car, n can take values between 0 and h. Finally, the finiteness of H ensures that for a sufficiently large value of p, n is zero. Within these restrictions, the function $n(z, p)$ can take any form. Hence, we may treat the function $n(z, p)$ as a 'primitive' as long as we keep the above restrictions in mind. In this chapter, we do precisely that and, in addition (this is a harmless and mathematically convenient assumption), ignore that fact that n and z can take only integer values.

Hence, from now on we shall treat n as a mapping (R being used to denote the set of all real numbers and R_+ all non-negative ones),

$$n: R \times R_+ \rightarrow [0, h],$$

which is continuous, monotonically non-decreasing in the first variable, non-increasing in the second variable, and for a sufficiently large value of the second variable n takes the value of zero. I shall refer to this as an n-function.

Given a price of $\hat{p}$ and a supply of $\hat{x}$, what are the equilibrium amounts that may be demanded? Before giving a formal answer, it is useful to have a diagrammatic representation.

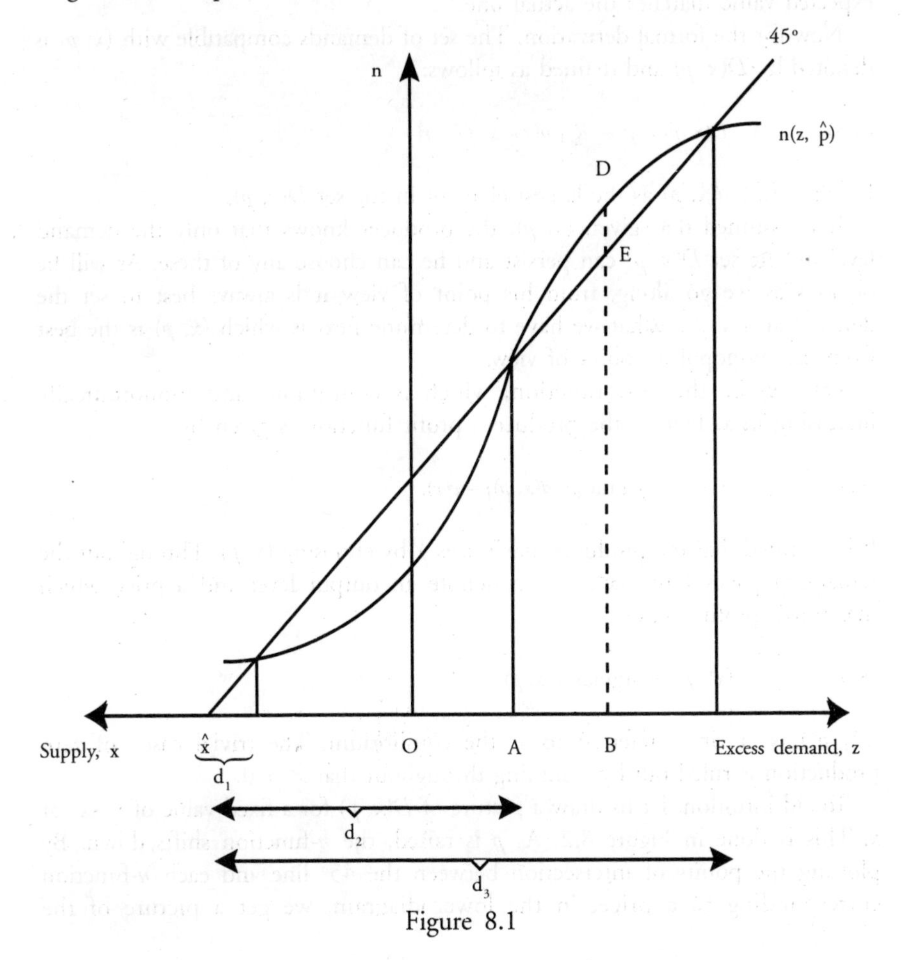

Figure 8.1

Figure 8.1 shows the function $n(z, \hat{p})$ for different values of z. The amount supplied is measured from the origin in the leftward direction and excess demand is measured in the rightward direction. In this case, the amount supplied is marked by $\hat{x}$. Given any level of demand d, to find out the amount of excess demand, we simply have to mark a distance of d from $\hat{x}$ (to the right of $\hat{x}$). Thus, if d_2 is demanded, the excess demand is equal to $0A$.

Now draw a 45° line through $\hat{x}$, as shown. Let us check whether a particular level of demand is compatible with $\hat{x}$ being supplied at price $\hat{p}$. Consider the demand given by the distance $\hat{x}B$ in the diagram. This implies an excess demand of $0B$, which in turn implies that aggregate demand will be BD. But BD is greater than BE, which equals $\hat{x}B$. Hence, demand equilibria are given by the points of intersection between the n-function and the 45° line. Therefore, demand levels of d_1, d_2, and d_3 are the only demands compatible with $(\hat{x}, \hat{p})$. This is an essential part of a rational-expectations equilibrium. According to this, an individual's valuation of the good depends not on excess demand, as in (8.1), but his *expectation of* excess demand. An equilibrium occurs when the expected value matches the actual one.[3]

Now for the formal derivation. The set of demands compatible with (x, p) is denoted by $D(x, p)$ and defined as follows:

$$D(x, p) = \{r \in R_+ \mid n(r - x, p) = r\} \tag{8.3}$$

Let us define $d(x, p)$ as the largest element in the set $D(x, p)$.

It is assumed that given (x, p), the producer knows that only the demand levels in the set $D(x, p)$ can persist and he can choose any of these. As will be obvious as we go along, from his point of view it is always best to set the demand at $d(x, p)$. What we have to determine next is which (x, p) is the best from the monopolist's point of view.

Let $c(x)$ be the cost function, which is continuous and monotonically increasing in x. Hence, the producer's profit function is given by

$$Y(x, p) = p \min \{x, d(x, p)\} - c(x). \tag{8.4}$$

It is assumed that the producer maximizes Y by choosing (x, p). Throughout the remaining pages I use (x^*, p^*) to denote an output level and a price which maximizes profit.[4] Thus

$$(x^*, p^*) = \text{argmax } Y(x, p) \tag{8.5}$$

(x^*, p^*) is at times referred to as the equilibrium. The trivial case of zero production is ruled out by assuming throughout that $x^* > 0$.

To aid intuition, let us draw a picture of $D(x, p)$ for a fixed value of x, say at $\hat{x}$. This is done in Figure 8.2. As p is raised, the n-function shifts down. By plotting the points of intersection between the 45° line and each n-function corresponding to a price, in the lower diagram, we get a picture of the

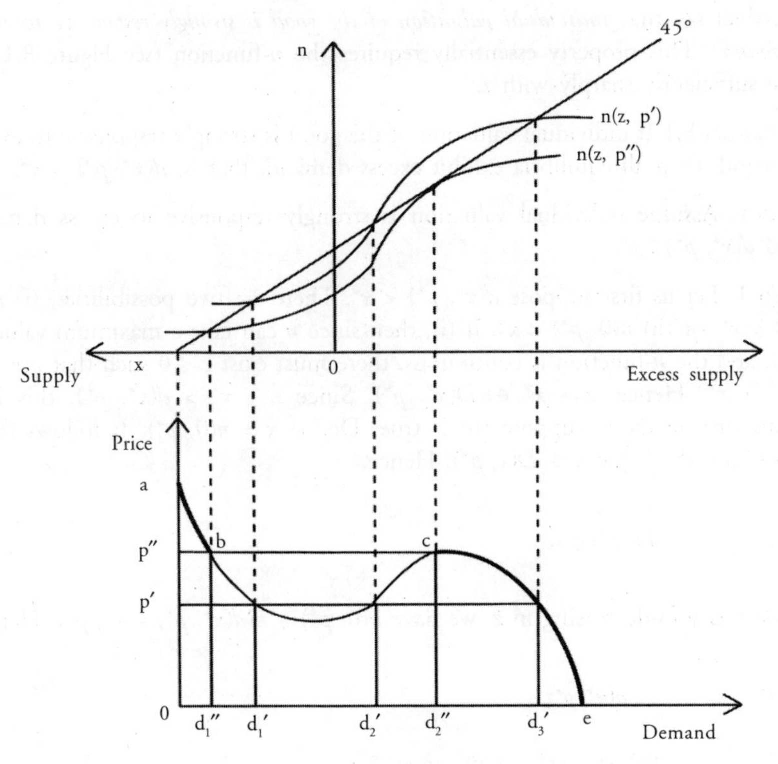

Figure 8.2

correspondence $D(\hat{x}, p)$ for different values of p. Thus, for example, $D(\hat{x}, p')=\{d_1', d_2', d_3'\}$. It is easy to see that $d(\hat{x}, p)$ plotted against p is the thick discontinuous line *abce*, which at p'' takes the value of c instead of b.[5]

The discontinuity is now clear. If, starting from a price of p'', the producer increases price a little, the maximum sustainable demand for his product drops from d_2''; to below d_1''. If, holding p'' constant, he raises supply to slightly above $\hat{x}$ (which will shift the 45° line up in Figure 8.2), a similar sharp decline in demand occurs. It is this kind of discontinuity which can explain why it may be profit-maximizing for a producer to maintain an excess demand for his product.

8.3. Excess-demand Equilibria

There are different ways of illustrating excess-demand equilibria. One is to construct an example. I shall, however, first use the more general—and hopefully more instructive—approach of establishing a sufficient condition for equilibria to exhibit excess demand. Later, an example is constructed to illustrate this.

For this we need a definition. Let us use $p(x)$ to denote the price (wherever such a price exists) which ensures $n(0, p(x)) = x$. If the n-function has the property for that for all $x > 0$ there exists $z > 0$ such that $n(z, p(x)) > x + z$, then we shall say that *individual valuation of the good is strongly responsive to excess demand.*[6] This property essentially requires the n-function (see Figure 8.1) to rise sufficiently sharply with z.

THEOREM 8.1 If individual valuation of the good is strongly responsive to excess demand, then all equilibria exhibit excess demand, that is, $d(x^*, p^*) > x^*$.

PROOF. Assume individual valuation is strongly responsive to excess demand and $d(x^*, p^*) \leqq x^*$.

Step 1. Let us first suppose $d(x^*, p^*) < x^*$. There are two possibilities: (i) $n(0, p^*) \geqq x^*$, or (ii) $n(0, p^*) < x^*$. If (i), then since n can take a maximum value of $\#H$, and the n-function is continuous, there must exist $\hat{z} \geqq 0$ such that $n(\hat{z}, p^*) = \hat{z} + x^*$. Hence, $\hat{z} + x^* \in D(x^*, p^*)$. Since $\hat{z} + x^* > d(x^*, p^*)$, this is a contradiction. Now suppose (ii) is true. Define $\hat{x} = n(0, p^*)$. It follows from definition (8.3) that $\hat{x} \in D(\hat{x}, p^*)$. Hence,

$$(8.6) \qquad d(\hat{x}, p^*) \geqq \hat{x}.$$

since n is non-decreasing in z, we have $n(0, p^*) \geqq n(d(x^*, p^*) - x^*, p^*)$. Hence,

$$(8.7) \qquad \hat{x} \geqq d(x^*, p^*),$$

$$\begin{aligned} Y(\hat{x}, p^*) &= p^*\hat{x} - c(\hat{x}) \text{ by (8.6)},\\ &\geqq p^*d(x^*, p^*) - c(\hat{x}) \text{ by (8.7)},\\ &> p^*d(x^*, p^*) - c(x^*) \text{ since } \hat{x} < x^*,\\ &= Y(x^*, p^*) \text{ since } d(x^*, p^*) < x^*. \end{aligned}$$

This contradicts the fact that (x^*, p^*) maximizes profit, and thereby establishes that $d(x^*, p^*)$ cannot be less than x^*.

Step 2. Suppose now that $d(x^*, p^*) = x^*$. Hence, $p(x^*) = p^*$. It follows from the fact that individual valuation of the good is strongly responsive to excess demand, that there exists $z > 0$, say z', such that $n(z', p^*) > x^* + z'$. By the continuity of n and the fact that n goes to zero for a sufficiently large p, it follows that there exists $p' > p^*$ such that

$$n(z', p') = x^* + z'.$$

This implies $x^* + z' \in D(x^*, p')$, by (8.3). Hence,

$$d(x^*, p') \geqq x^* + z'.$$

Therefore,

$$\begin{aligned} Y(x^*, p') &= p'x^* - c(x^*) \text{ since } z' > 0,\\ &> p^*x^* - c(x^*),\\ &= Y(x^*, p^*) \text{ since } x^* = d(x^*, p^*). \end{aligned}$$

This contradiction establishes that $d(x^*, p^*)$ is not equal to x^*.

Together with Step 1, this establishes a contradiction of our initial assumption that $d(x^*, p^*) \leqq x^*$.

The remainder of this section is used to construct a numerical example to illustrate a monopoly equilibrium which exhibits excess demand, that is, the kind of situation described in the above theorem.

Suppose the n-function is given as follows:

$$n(z, p) = 100 - 10p \quad \text{for all } z \geqq 20, \tag{8.8}$$
$$= \max\{0, 60 - 10p + 2z\} \quad \text{for all } z < 20.$$

Thus if price is 5 and excess demand is 25, the demand for this good will be 50. Assume, further, that this good can be produced at zero cost by the monopolist. That is, $c(x) = 0$, for all x.

It will now be shown that, at equilibrium, the price of the good will be 4, supply 40, and demand 60; thereby implying an excess demand of 20.

Let us first derive the function $d(x, p)$ from equation (8.8).

$$d(x, p) = 100 - 10p \quad \text{for all } p \leqq 8 - (x/10), \tag{8.9}$$
$$= 0 \quad \text{for all } p > 8 - (x/10).$$

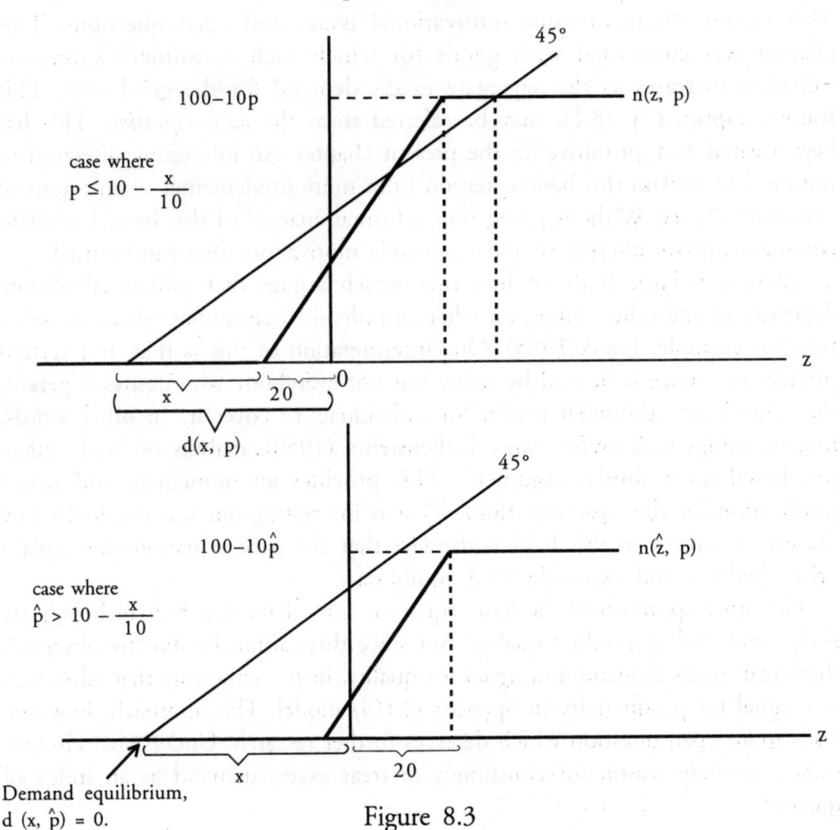

Figure 8.3

Figure 8.3 depicts the derivation of (8.9). Recall that the points of intersection between an $n(z, p)$ curve and the 45° line through x depict equilibrium levels of demand for a given x and p, and the largest of these demands is $d(x, p)$.

Since $p \leqq 8 - (x/10)$ implies $\min \{x, d(x, p)\} = x$, and since $c(x) = 0$, we get

$$Y(x, p) = 0, \quad \text{for all } p > 8 - (x/10),$$
$$= px \quad \text{for all } p \leqq 8 - (x/10).$$

For each value of x, $Y(x, p)$ is maximized by setting

$$p = 8 - (x/10). \tag{8.10}$$

It follows that to maximize $Y(x, p)$, we have to maximize $(8 - (x/10))x$. The first-order condition for achieving this is

$$8 - (x/5) = 0.$$

Hence, $x = 40$. By inserting this in (8.10), we get $p = 4$. Equation (8.9) implies $d(40, 4) = 60$. This establishes that at equilibrium there is an excess demand of 20. The monopolist cannot respond to this by raising price because this would cause demand to plummet so much that he would be worse off in the end.

8.4. Extensions

This section discusses some motivational issues and open questions. This chapter was concerned with goods for which each consumer's subjective valuation increases as the aggregate excess demand for the good rises. This feature, captured in (8.1), may be referred to as the *basic equation*. This has been treated as a 'primitive' in the present chapter. An interesting direction to pursue is to explain this basic equation from more fundamental assumptions of consumer theory. Without going into a formal analysis of this here, I want to comment on two alternative and reasonable motivations that can be used.

There is a large body of literature which argues that individual welfare depends, among other things, on what an individual consumes *relative to others* (see, for example, Frank 1985). One interpretation of this is that, if a certain product or service is desired by many but not everybody who desires it gets it, then this is an additional reason for individuals to covet it. In other words, human beings seek *exclusiveness*. Leibenstein's (1950) analysis of 'snob' effect was based on a similar argument. This provides an immediate and direct justification for the basic equation. What is interesting but was overlooked by the earlier writers in this field is the fact that the basic equation can explain price rigidities and excess-demand equilibria.

The other approach to the basic equation is to claim that human beings are really interested in product quality. But since this cannot be directly observed, they treat excess demand as a *signal* for quality, in the same way that education is a signal for productivity in Spence's (1974) model. This approach, however, leads to an open question which deserves further research: Under what circumstances is there reason for consumers to treat excess demand as an index of quality?

Turning to another problem concerning primitives, note that in this chapter we do not begin from utility functions but by directly specifying each person *i*'s behavioural function (8.1), which is later aggregated to (8.2). Can these be derived from standard utility maximization? The answer to this is yes and this is demonstrated by a utility function which belongs to a class of functions which has been used widely in recent years (see, for example., Gabszewicz and Thisse 1979; and Shaked and Sutton 1983).

Suppose that a consumer $t \in [0, h]$ has an income of $W(t) = W_1 + W_2 t$, where W_1 and $W_2 (\geqq 0)$ are exogenously given parameters. The utility that consumer t gets by consuming a unit of the good, characterized by price p and excess demand z, is $u(z)(W(t) - p)$, where $u(z)$ is a non-decreasing function of z. The utility that he gets if he does not consume the good is given by $W(t)u_0$. It is easy to check that the number of people demanding the good, characterized by (p, z), is[7]

$$n(z, p) = h - u(z)p/(u(z) - u_0) + W_1/W_2.$$

So what was earlier treated as a primitive, to wit, the n-function, is now derived from a standard utility function.[8] Of course, the important question as to how the function $u(z)$ is, in turn, derived is not answered here.

Finally, a comment on labour markets. There has been a large recent literature on 'efficiency' wages (for example, Shapiro and Stiglitz 1984), which explains downward wage rigidity in terms of the *employer's* preference for not lowering wages. The model of this chapter can be used to construct a new efficiency-wage argument.

Consider a person seeking a job and evaluating alternative possibilities. There is a lot of sociological evidence (Jencks et al. 1985) that a person will be concerned about, among other things, the social status of a potential job. And one way of judging the status associated with a firm is by the excess supply of employees faced by it. Admittedly this argument would be more applicable to salaried and higher-paid jobs, in contrast to several existing theories which apply mainly to the low-wage sector (for example, Mirrlees 1975). But this complementarity may well be its advantage.

NOTES

1. It has, for instance, been common practice for companies liquidating their foreign holdings and selling shares to Indian citizens in response to the Foreign Exchange Regulation Act to price them so as to have a large excess demand. The same has been observed for debenture issues of private companies.
2. It is worth noting that a more sophisticated theory would require v_i to be a function of z and p, i.e., $v_i = v_i(z, p)$, instead of (8.1). But this complication is inessential, given the problem addressed in this chapter. It will be clear as we go along that as long as aggregate excess demand figures in the individual demand function—no matter in what form—the model developed in this chapter will be relevant. Hence there is no harm in using the simple idea captured in (8.1).
3. A similar idea is used and elaborated upon in Basu (1986).

4. The existence of a maximum is guaranteed by the following argument. Let P be a price such that for all prices above this Y is non-positive. The existence of such a P follows from the properties of the n-function. Clearly there is no advantage in raising x to above h. Hence x may be taken as varying between 0 and h. Hence the domain of the function, $Y(\cdot)$, is $[0, h] \times [0, P]$, which is compact. It is easy to check that $Y(\cdot)$ is upper semi-continuous, thereby ensuring that it achieves a maximum somewhere in the domain (Berge 1963: 74).
5. It is interesting to note that this demand curve, $d(x, p)$, is downward sloping in p. However, the proof is very different from the one used to establish the downward slope of the textbook demand curve. For a proof suppose $p^0 < p^1$. We have to show that $d(x, p^1) < d(x, p^0)$.
   Define $z^0 \equiv d(x, p^0) - x$. Since $n(\cdot)$ is non-decreasing in p,
   $$n(z^0, p^1) \leqq n(z^0, p^0).$$
   Next, it is proved that
   (i) for all $z > z^0$, $n(z, p^0) < x + z$.
   Let $z' > z^0$ and suppose
   (ii) $n(z', p^0) \geqq x + z'$.
   Since $n(\cdot)$ is bounded, there must exist $\hat{z} > z'$ such that
   (iii) $n(\hat{z}, p^0) < x + \hat{z}$.
   (ii), (iii) and the continuity of $n(\cdot)$ imply that there exists $\ddot{z} \in [z', \hat{z}]$ such that $n(\ddot{z}, p^0) = x + \ddot{z}$. This implies that
   $$d(x, p^0) \geqq x + \ddot{z}$$
   $$> x + z^0 = d(x, p^0).$$
   This is a contradiction. Hence, (i) must be true, which implies that, for all $z > z^0$, $n(z, p^1) < x + z$. Hence, $d(x, p^1) \leqq x + z^0 = d(x, p^0)$.
6. The choice of terminology is clear once it is appreciated that this property is equivalent to requiring that for a 'large' number of people v_i rises sharply in response to an increase in excess demand.
7. Given that the number of people demanding the good cannot be less than 0 or more than h, a more fastidious specification of the n-function is as follows:
   $$n(z, p) = \text{mid}\ \{0, h - u(z)p/(u(z) - u_0) + W_1/W_2, h\},$$
   where mid $\{a, b, c\}$ is a number which is second largest (ties being broken arbitrarily) among a, b, and c.
8. Further, $n(z, p)$ is concave in z if $u(\cdot)$ is concave.

REFERENCES

Basu, K. 1986. 'The Market for Land: An Analysis of Interim Transactions'. *Journal of Development Economics* 20.

Berge, C. 1963. *Topological Spaces*. Edinburgh: Oliver and Boyd.

Frank, R.H. 1985. *Choosing the Right Pond: Human Behaviour and the Quest for Status*. Oxford: Oxford University Press.

Gabszewicz, J.J. and J.-F. Thisse. 1979. 'Price Competition, Quality and Income Disparities'. *Journal of Economic Theory* 20.

Jencks, C., L. Perman, and L. Rainwater. 1988. 'What is a Good Job? A New Measure of Labour Market Success'. *American Journal of Sociology* 93 (May): 1322–57.

Leibenstein, H. 1950. 'Bandwagon, Snob and Veblen Effects in the Theory of Consumers' Demand'. *Quarterly Journal of Economics* 64.

Mirrlees, J.A. 1975. 'Pure Theory of Underdeveloped Economies'. In L.G. Reynolds, ed. *Agriculture in Development Theory*. New Haven, CT: Yale University Press.

Scitovsky, T. 1944. 'Some Consequences of the Habit of Judging Quality by Price'. In T. Scitovsky, *Papers on Welfare and Growth.* Stanford CA: Stanford University Press.

Shaked, A. and J. Sutton. 1983. 'Natural Oligopolies'. *Econometrica* 51.

Shapiro, C. and J. Stiglitz. 1984. 'Equilibrium Unemployment as a Worker Discipline Device'. *American Economic Review* 74.

Spence, A.M. 1974. *Market Signalling.* Cambridge, MA: Harvard University Press.

9 Why Monopolists Prefer to Make their Goods Less Durable

9.1. Introduction

This chapter is concerned with consumer goods that depreciate over time, such as footwear, clothes, carpets, and wallpaper. With any of these goods the consumer has to decide, at some point, whether to continue to use it or to discard it and buy afresh. This, in turn, presents the producer with a problem: how durable should he make his product? The relation between market structure and durability has been the subject of a considerable debate.[1] The existing literature, however, ignores an important feature of durable goods—the fact that their durability can be used as a screening device.

At first glance, it seems that increasing durability would reduce profit by reducing the frequency of purchase. This was Chamberlin's (1957) view; but it is flawed because, as the producer raises durability, he can raise price to compensate for the reduction in sales (see, for example, Barro 1972). It seems, therefore (and this belief is widely shared in the literature), that a profit-maximizing monopolist would raise durability as much as possible, if production costs were independent of durability and the capital market were perfect. It is shown in this chapter that this position is, in turn, untenable once we recognize that consumer preferences are heterogeneous.

This is formalized here by assuming that there are two types of consumers: the 'lavish', who are fastidious about quality and would buy a new unit as soon as the old one is slightly worn out, and the 'thrifty', who prefer to use goods up to a point of greater depreciation. Now, even if a monopolist has to charge the same price from each consumer, he can discriminate between the two groups by choosing a suitable amount of durability. For instance, if he makes the good

From *Economica* 55, 1998: 541–6.

A first version of this chapter was written at the Institute for Advanced Study, Princeton. I am grateful to Avinash Dixit, Michael Katz, Ekkehart Schlicht, and Nirvikar Singh for comments.

less durable, the lavish consumers make more frequent purchases per unit of time. Hence, although both consumer types pay the same price per unit of *commodity*, the lavish ones pay more per unit of *time*. That is, they pay a higher *effective* price. Hence, by making the product less durable, the monopolist manages to price-discriminate between consumers.

8.2. Durability and Consumer Behaviour

I am concerned with a consumer good that depreciates over time. It is not a good that remains new for n days and then disintegrates but, more realistically, one that loses its newness with time before finally disintegrating.[2] The simplest way to capture this idea is to assume that the good lasts one period (and then completely disintegrates), but within this period, during the first q units of time the good is new and has a quality level of N units, and during the remaining (1-q) units of time it is depreciated and has a quality level of D units. Of course, $N > D$.

The extent of durability of the good is captured by this parameter q. If q becomes larger, we say that the good is now 'more durable' or that it 'depreciates more slowly'. If $q = 1$, the good is described as 'fully durable'. If $q = 0$, then this good is never new. We are here interested in goods that are new for some time, however little. Hence we assume that there exists a very small real number $e > 0$ such that q must be greater than or equal to e. Throughout, it is assumed that q is a variable which the producer has to select from the interval $[e, 1]$. But before we go on to the question of producer's decision, we need to model the consumer.

For simplicity, I shall suppose that we are talking of a good which, by its very nature, is such that the consumer would want to possess at most one unit at each point of time. Toothbrushes, pianos, dining tables, and wallpaper are, examples of such goods.[3] Hence, the consumer's main problem is whether to buy the good, and if 'yes', how often to replace it (i.e., discard it and buy a new one).

Assume that there are n consumers, indexed by i. Let $Q(t)$ be the density function of the number of units of quality that consumer i gets over time, and let M_i be his income and X_i his expenditure on this good in one period. Hence, he spends $M_i - X_i$ units of money on other goods. The utility that consumer i gets from one period is assumed to be given by

$$U^i = \int_{t=0}^{1} f^i(Q(t))\, dt + M_i - X_i$$

where $\partial f^i / \partial Q > 0$. Clearly we can redefine the function f^i so as to include M_i. Hence, without loss of generality, we write i's utility function as

$$U^i = \int_{t=0}^{1} f^i\,(Q(t))dt - X_i$$

where $\partial f^i / \partial Q > 0$ and $f^i(0) = 0$. Note that, since each period looks the same, U^i may be interpreted as the average utility over steady state.

Let us compute the utilities associated with three specific options. First, if the consumer does not buy this good, $Q(t) = 0$ for all t (it is being assumed that quality is something that only this good provides) and $X_i = 0$. The utility that he gets by not buying the good is denoted by U^{i0} and is given by

$$U^{i0} = 0$$

Second, let U^{iL} be the utility that he gets if he decides to buy afresh as soon as the old good depreciates in quality. Denoting the price of the good by p, it is obvious that

$$U^{iL}(p, q) = f^i(N) - p/q. \tag{9.1}$$

If a person buys afresh as soon as the good depreciates, we shall say that he acts *lavishly*.[4]

Third, let U^{iT} be the utility he gets if he chooses to buy one unit at the beginning of the period and not buy again in this period; that is, he acts *thriftily*. Clearly,

$$U^{iT}(p, q) = f^i(N)q + f^i(D)(1 - q) - p. \tag{9.2}$$

We have so far considered three options: acting lavishly, acting thriftily, and not buying. There are other options, like buying one unit, after it depreciates continuing with it for some time and then buying a new one again. It is easy to check that such options can be ignored because the consumer's best strategy always includes one of: not buying, being thrifty, and being lavish. Given this, the following tie-breaking assumption is fairly innocuous and is maintained throughout this chapter: faced with a (p, q), if a consumer is indifferent between buying and not buying a good, he buys it; and, having decided to buy, if he is indifferent between acting thriftily and lavishly, he acts lavishly.

9.3. Monopoly

Suppose there is one producer and he offers the price–durability pair (p, q). Let c be the cost of producing each unit of good. It is assumed for simplicity that the cost does not depend on the product's durability.[5] Note that this is the most adverse assumption for what I am trying to establish.

If $n^L(p, q)$ and $n^T(p, q)$ are the numbers of consumers who act lavishly and thriftily, then the monopolist's profit, R, is given by

$$R(p, q) = (p - c)\left\{\frac{n^L(p, q)}{q} + n^T(p, q)\right\}. \tag{9.3}$$

Definition. 9.1(p^*, q^*) is a monopoly equilibrium if and only if

$$(p^*, q^*) = \operatorname{argmax} R(p, q)$$

with $p \in [0, \infty)$, $q \in [e, 1]$.

It will be assumed throughout that there exists some (p, q), that gives the monopolist a positive profit.

It will now be shown that, although it does not cost the monopolist anything to make the product more durable, he may nevertheless prefer to make it less durable (i.e., to set $q < 1$). For this result and other discussions in this section, I make use of the fact that a monopoly equilibrium always exists. Hence, we first need to establish existence.

Let P be a price at which consumer demand is 0. Define a set K as follows:

$$K = \{(p, q) \mid p \in [c, P], q \in [e, 1]\}.$$

Consider the real-valued function R on K. It may be checked that R is an upper semi-continuous function. This follows from our tie-breaking assumption in Section 9.2 which says that, whenever a consumer is indifferent between any two of the options in the set {not buy, act lavishly, act thriftily}, he will choose the one that gives the producer a larger profit. Since K is compact, it follows that R attains a maximum in K. Since for all (p, q) not belonging to K, $R(p, q) \leq 0$ and we have assumed that for some (p, q), $R(p, q)$ must be positive, it follows that the maximum value attained by R in K is greater than zero. Hence that is also the maximum value that R attains on an unrestricted domain of price-durability pairs, that is, with $p \in [0, \infty)$ and $q \in [e, 1]$. The existence of monopoly equilibrium being assured, we may now examine the relation between monopoly and durability.

THEOREM 9.1: There exist equilibria where monopolists sell goods of limited durability (i.e., $q < 1$). A necessary condition for this to happen is that consumers have heterogeneous preference.

PROOF: The following example establishes the first part of the theorem. Assume there are only two consumers, with the following characteristics:

$$f^1(N) = 4; \quad f^1(D) = 0$$

$$f^2(N) = 2; \quad f^2(D) = 1$$

In addition, assume $c = 0$ and $e \leq 1/4$. Using equations (9.1) and (9.2), we get

$$\text{(9.4)} \quad \begin{cases} U^{1L}(p, q) = 4 - p / q; & U^{1T}(p, q) = 4q - p \\ U^{2L}(p, q) = 2 - p / q; & U^{2T}(p, q) = q + 1 - p. \end{cases}$$

Let us check the maximum profit that the monopolist can earn, given that q is fixed at 1 (that is, we want the value of max $R(p, 1)$). Inserting $q = 1$ in (9.4), it is clear that as long as $p \leq 2$ he can sell two units (one each to consumers 1 and 2), and if p is in [2, 4] he can sell one unit (to consumer 1). Hence, max $R(p, 1) = R(4, 1) = R(2, 1) = 4$. To prove the theorem, we merely have to show that there exists q in the interval $[e, 1]$ such that, for some p in $[0, \infty)$, $R(p, q) > 4$. Consider $q = 1/4$ and $p = 1$. Inserting these in (9.4), we get

$$U^{1L} = 0; \quad U^{1T} = 0$$

$$U^{2L} = -2; \quad U^{2T} = 1/4.$$

Hence 1 will act lavishly and 2 thriftily. That is, 1 will buy four $(=1/q)$ units and

2 will buy one unit. Hence, $R(1, ¼) = 5$. Since we know that a monopoly equilibrium always exists, the equilibrium value of q must be less than 1.

In order to prove the second part, suppose $f^1 = \ldots = f^n$, (p^*, q^*) is a monopoly equilibrium and $q^* < 1$. Since consumers are identical, either (i) everybody acts lavishly or (ii) everybody acts thriftily. Suppose (i): then

$$R(p^*, q^*) = (p^* - c)\frac{n}{q^*} \quad (9.5)$$

and from (9.1) and (9.2) it follows that (since $U^{iL} \geq U^{iT}$)

$$f^i(N) - f^i(D) \geq p^*/q^*, \text{ for all } i. \quad (9.6)$$

If the monopolist raises p and q proportionately (that is, ensuring that $p/q = p^*/q^*$), then (9.6) implies that everybody remains lavish and (9.5) implies an increase in profit.

Suppose (ii). Then choose $p > p^*$ and $q > q^*$ so that

$$U^{iT}(p, q) = U^{iT}(p^*, q^*).$$

Two things can happen with this: (a) people remain thrifty, or (b) people become lavish. If (a), then

$$R(p, q) = (p - c)n > (p^* - c)n = R(p^*, q^*).$$

If (b), then

$$R(p, q) = (p - c)\frac{n}{q} > (p^* - c)n = R(p^*, q^*).$$

This contradiction establishes that the original situation could not have been an equilibrium. Q.E.D

There are several models (for example, Parks 1974) in which monopolists offer many price–durability (or price–quality) pairs and practise price discrimination. In such models it is expected that monopolists would be able to screen consumers because the problem is analogous to the screening in standard non-linear pricing and optimal income tax theory.[6] What is interesting in the present model is that, even by offering a *unique* price–durability pair, the monopolist manages to price-discriminate. He does this by inducing different consumers to respond differently and thereby to pay different *effective* prices for the good.

Finally, note that, while the question of durability is obviously related to the one of quality, my central theorem exploits the natural relation between *durability* and the frequency of purchase. Hence, this problem is distinct from the problem of quality choice as discussed in, for instance, Sheshinski (1976) and Mussa and Rosen (1978).

9.4. Competition

In the case of competition with identical firms and free entry, products would invariably be fully durable.[7] This is because in such a case profits of firms would be 0. Now if $q < 1$, a firm can raise q a bit and also raise p, but sufficiently little to retain some customers. It will then be earning a positive profit since p will exceed c. This contradiction establishes that $q = 1$.

In concluding, it may be pointed out that it should be possible to use this model to throw light on the relation between durability and second-hand markets, the effect on durability of intermediate market structures like oligopoly, and the engineering of fashions and their decay.

NOTES

1. See, for example, Levhari and Srinivasan (1969); Sieper and Swan (1973); Kamien and Schwartz (1974); Leibowitz (1982).
2. This kind of assumption is used by, among others, Leibowitz (1982) and Bond and Samuelson (1984). Stokey's (1981) and Bulow's (1982) durable goods, on the other hand, never lose their newness.
3. Such goods have been analysed by several authors; see, for example, Gabszewicz and Thisse (1979). In the present model this assumption can be relaxed at the expense of a messier algebra.
4. We ignore the problem that p / q might not be an integer.
5. Several authors in this area make this same assumption: see, for example, Barro (1972).
6. It is easy to demonstrate that in some situations a monopolist will, given a choice, prefer to offer more than one (p, q) pair and to offer at least one less durable product. Consider the example where $f^1(N) = 5, f^1(D) = 0, f^2(N) = 2, F^2(D) = 1, e = 1/8$ and $c = 1$. If the monopolist has to offer one (p, q) pair, he would choose (5, 1) and his profit would be 4. It is easy to see that, if he were allowed to offer more than one type of good, he could earn a larger profit by offering, for example, the following (p, q) pairs: (5,1) and ($1\frac{1}{6}$, 1/6). Person 1 would choose the former and 2 the latter, and profit would be equal to 4 plus 1/6.
7. A formal model of this section is available from the author on request.

REFERENCES

Barro, R.J. 1972. 'Monopoly and Contrived Depreciation'. *Journal of Political Economy* 80: 598–602.

Bond, E.W. and L. Samuelson. 1984. 'Durable Good Monopolies with Rational Expectations and Replacement Sales'. *Rand Journal of Economics* 15: 336–45.

Bulow, J.I. 1982. 'Durable-goods Monopolists'. *Journal of Political Economy* 90: 314–32.

Chamberlin, E.H. 1957. *Towards a More General Theory of Value.* New York: Oxford University Press.

Gabszewicz, J.J. and J.F. Thisse. 1979. 'Price Competition, Quality and Income Disparities. *Journal of Economic Theory* 20: 340–59.

Kamien, M.I. and N.L. Schwartz 1974. 'Product Durability Under Monopoly and Competition'. *Econometrica* 42: 289–301.

Leibowitz, S.J. 1982. 'Durability, Market Structure, and New-used Goods Models'. *American Economic Review* 72: 816–24.

Levhari, D. and T.N. Srinivasan. 1969. 'Durability of Consumption Goods: Competition versus Monopoly'. *American Economic Review* 59: 102–7.

Mussa, M. and S. Rosen. 1978. 'Monopoly and Product Quality'. *Journal of Economic Theory* 18: 301–7.

Parks, R.W. 1974. 'The Demand and Supply of Durable Goods and Durability'. *American Economic Review* 64: 37–55.

Sheshinski, E. 1976. 'Price, Quality and Quantity Regulation in Monopoly Situations'. *Economica* 43: 127–37.

Sieper, E. and P.L. Swan 1973. 'Monopoly and Competition in the Market for Durable Goods'. *Review of Economic Studies* 40: 333–51.

Stokey, N.L. 1981. 'Rational Expectations and Durable Goods Pricing'. *Bell Journal of Economics* 12: 112–28.

10 Entry-Deterrence in Stackelberg Perfect Equilibria

with Nirvikar Singh

10.1. Introduction

As had been pointed out by Bain (1956) and Sylos-Labini (1962) and is well known now, the behaviour of a firm depends as much on its existing rivals as on the potential ones. This is the essence to analysing barriers to entry. The present chapter analyses entry-deterrence in a duopoly, consisting of an incumbent and an entrant, where the post-entry game is Stackelbrg with the incumbent playing leader. While a number of alternative characterizations of the post-entry game have been discussed in the literature (e.g., Cournot-Nash by Dixit 1980; Stackelberg with entrant as leader by Salop 1979), this particular characterization has been ignored as uninteresting (Saloner 1985, being an exception). And indeed it would be uninteresting if the standard cost-function (for example, the kind used by Spence 1977; Dixit 1979, 1980) is used. In contrast, we consider a more realistic cost function in which fixed costs consist of two parts: *entry cost* and (production-) *commencement cost*. The former is the cost associated with entering an industry (acquiring a licence, which in turn may require setting up an office, lawyer fees, etc.) and the latter is the usual cost of beginning positive production.

In this chapter we also allow firms to go in for a broader range of commitments than is allowed for in the literature.[1] With these modifications the model of duopoly with the incumbent playing leader in the post-entry game

From *International Economic Review* 31 (1, February), 1990: 61–71.

Financial support from the UCSC Faculty Research Committee is gratefully acknowledged by Nirvikar Singh. A part of the work on this chapter was done while Kaushik Basu was a Member at the Institute for Advanced Study, Princeton. The authors have benefited from comments received at a seminar at the Indian Statistical Institute, Delhi.

(which is after all more natural than the entrant playing leader) becomes an interesting case explaining a large class of phenomena.

10.2. The Framework

Let x_1 and x_2 be the sales of firms 1 and 2 and let the inverse demand function be $p(x_1 + x_2)$ that is, the firms produce a homogeneous good. We assume (A10.1) p is thrice-differentiable, $p' \leq 0$ and there exists a real number n such that $p(n) = 0$.

We shall use 1 and 2 to represent, respectively, the incumbent and the entrant firms. It is also useful to stress the difference between sales and production, because firms might want to build up stocks, and x_1 throughout represents firm *i*'s sales, rather than output. The incumbent firm is able to pre-commit in two ways. These variables are denoted k_1 and r_1 respectively, and they enter the cost function of firm 1 as follows:

$$c_1(r_1, k_1, x_1) = f_1 + (v_1 - r_1)\, x_1 + r_1 \max\{k_1, x_1\}. \tag{10.1}$$

If r_1 were exogenous, this would be the standard cost function (the one used by Dixit 1980; Bulow et al. 1985; and others) with r_1 being the cost of capital and v_1 the per unit cost of all inputs. However, we allow firm 1 to choose r_1, to represent the fact that incumbent firms may pre-commit a given level of potential output (k_1) with varying degrees of readiness (r_1). This is a realistic assumption. The standard practice of treating r_1 as fixed seems unduly restrictive. After all, there is no reason why a firm cannot actually undertake a certain

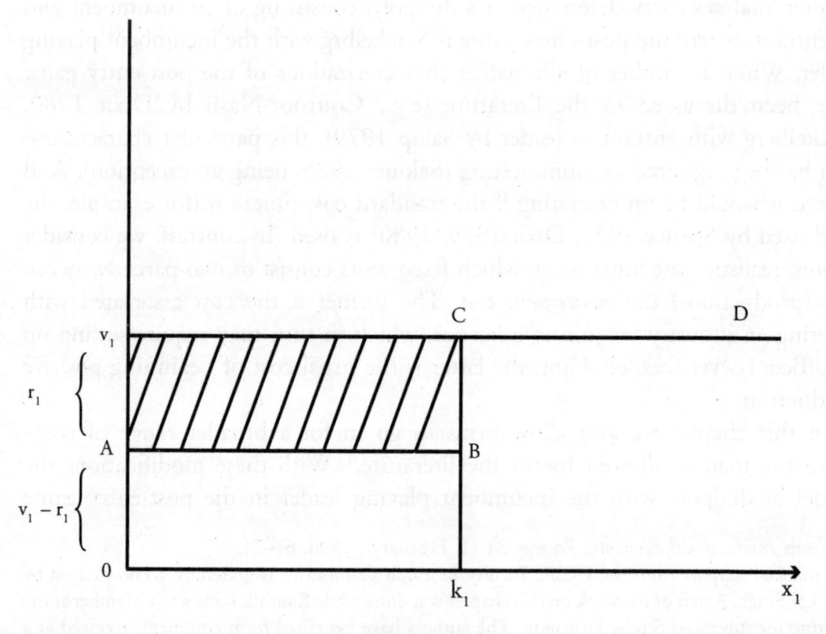

Figure 10.1

level of production and hold a part of it as stocks to threaten a potential entrant. This means that $r_1 = v_1$ is always open to firms.[2] We go further in treating *rl* as a variable which can take any value within $[0, v_1]$. Of course, we are measuring inputs in units of output.

One could get an intuitive picture of the kind of pre-commitment we are modelling by considering the marginal cost curve implied by (10.1). This is illustrated in Figure 10.1. If the firm pre-commits nothing ($r_1 = 0$, $k_1 = 0$) then the marginal cost curve is given by the line v_1D. If, however, $r_1 > 0$, $k_1 > 0$ (the case illustrated in the figure), then the shaded area is the cost that is pre-committed (that is, effectively this represents a 'fixed' cost). Hence, the *effective* marginal cost curve is given by the line ABCD, which is exactly the case considered by Dixit, excepting for the fact that the height of the line segment AB is, in the present model, an endogenous variable. Note also that x_1 may exceed k_1. It would generally be the case that $f_1 > 0$, but in the case of firm 1 the existence of fixed costs makes no difference. So, for simplicity, we set $f_1 = 0$. Finally, note that the (piecewise) linear cost specification is restrictive, but we shall see that it is sufficiently rich to encompass a wide range of possible equilibria.

Next, firm 2's cost function is

$$c_2(x_2) = \begin{cases} 0 & \text{if } x_2 = 0, \\ f_2 + v_2 x_2 & \text{if } x_2 > 0. \end{cases} \tag{10.2}$$

Hence, we are interpreting the fixed cost, f_2, as a cost which is incurred only if firm 2 enters and produces. The sales of firm 2 are denoted by x_2 (but since for firm 2 sales and production never differ, x_2 may be treated as either). It would be possible to introduce fixed costs associated with entry itself. This is inessential: the crucial requirement is that there be some fixed cost associated with positive production.

A clarifying comment on our cost function: Using D as a variable which takes a value of 1 if firm 2 enters the industry and 0 if it does not, a more general cost function would be as follows:

$$\begin{aligned} c_2(x_2, D) &= 0, \text{ if } D = 0 \\ &= K, \text{ if } D = 1, \text{ if } x_2 = 0 \\ &= K + f_2 + v_2 x_2, \text{ if } x_2 > 0. \end{aligned}$$

(It is assumed throughout that $D = 0$ implies $x_2 = 0$.) Implicit in Dixit's and Spence's work is the assumption that $f_2 = 0$, but $K > 0$. Our chapter, on the other hand, makes crucial use of the fact that $f_2 > 0$.[3] In fact, the case of the post-entry game being Stackelberg turns out to be interesting precisely because we do not assume away such fixed costs. In reality of course K would also be positive. This plays no role in our model and hence we set $K = 0$, which immediately reduces the cost function above to (10.2). So 'entry' in this chapter is a costless acquisition of a licence to produce. The essential consequence of a positive *entry* cost (that is, $K > 0$) is introduced in our chapter by the simple lexicographic assumption that if a firm earns zero profit after entry, it would prefer not to enter. We could have, at the cost of some additional algebra, instead assumed that $K > 0$.

Given the above duopoly situation, we assume that firm 1 first chooses (r_1, k_1), its pre-commitment. Then firm 2 decides whether to enter or not, that is, whether to set $D = 1$ or 0. After firm 2 chooses D, firm 1 selects its sales level x_1, followed by firm 2's choice of x_2. Hence an *outcome* of this sequence of choices is denoted by the quintuple (r_1, k_1, D, x_1, x_2). Since, in the post-entry game with $D = 1$, firm 1 chooses first, it is a Stackelberg leader and 2 a Stackelberg follower.

We introduce another assumption and some more notation before describing formally our equilibrium concept. A convenient simplification is that the potential entrant's output, given the incumbent's decisions, is unique. This has two parts, embodied in the following: Let $\Pi_2(x_1, x_2) = x_2\, p(x_1 + x_2) - c_2(x_2)$, i.e., Π_2 (.) is firm 2's profit function.

(A10.2) (i) $\Pi_2(x_1, x_2)$ is strictly concave in x_2.

(ii) Firm 2 will not produce (enter) if it can at best earn the same profit by producing (entering) as by not producing (entering).

A10.2 (i) ensures that firm 2's response to firm 1's sales is a function, say $R(x_1)$, except at the discontinuity created by $f_2 > 0$. By A 10.2 (ii), $R(x_1)$ is 0 at that level of x_1. It is easy to check that $R(x_1)$ is also 0 for greater x_1. We may define B_1 to be the smallest sales of firm 1 for which firm 2 produces nothing:

DEFINITION 10.1: $B_1 \equiv \min \{x_1 \mid R(x_1) = 0\}$.

We may also note that A10.2 (i) may be derived from standard assumptions on the first two derivatives of the inverse demand (see, for example, Friedman 1977).

In the next section, we begin analysis of the equilibrium outcomes of the duopoly situation above. The equilibrium notion we use is that of a perfect equilibrium. We therefore assume that at each move each firm chooses so as to maximize its profit, and takes into account that in the subsequent moves everyone will choose so as to maximize profits. An outcome which emerges from such a sequence of decisions will be referred to as a *Stackelberg perfect equilibrium,* or *SP-equilibrium* in brief.

10.3. Basic Results

In this and subsequent sections, we are chiefly concerned with the behaviour of firm 1, since firm 2 is a Stackelberg follower. Hence we drop the '1' subscript for the incumbent where it is unambiguous. Firm 1's profit is described by

$$\Pi_1(r_1, k_1, D, x_1, x_2) = \begin{cases} p(x_1 + x_2)x_1 - c_1(r_1, k_1, x_1), & \text{if } D = 1 \\ p(x_1)x_1 - c_1(r_1, k_1, x_1), & \text{if } D = 0. \end{cases} \tag{10.3}$$

If $D = 1$, then $x_2 = R(x_1)$ and we define $\tilde{\Pi}_1 (r_1, k_1, 1, x_1) \equiv \Pi_1 (r_1, k_1, 1, x_1, R(x_1))$. Corresponding to A10.2 (i), then, we have the following assumptions on Π_1:

(A10.3) (i) $\Pi_1 (r_1, k_1, 0, x_1, 0)$ is strictly concave in x_1.

(ii) $\tilde{\Pi}_1 (r_1, k_1, 1, x_1)$ is strictly concave in x_1.

A10.3(i) is again a standard assumption, but A10.3(ii) involves the third derivative of the inverse demand. A more explicit condition could be derived, but is not particularly insightful. A linear demand curve satisfies all these assumptions, for example (for the cost functions used here). In what follows, including the lemmas and theorems, A10.1, A10.2, and A10.3 are assumed throughout.

We now define different profit-maximizing levels of output for firm 1. These exist and are unique by our assumptions. In the following, 'argmax' denotes the maximizing value.

DEFINITION 10. 2: $\phi (r) = \text{argmax}_x\ \theta (x, r)$ with $\theta (x, r) \equiv p(x)x - (v - r)x$.

Hence $\phi (r)$ denotes the output level that a monopolist would choose if his marginal cost of production was $(v - r)$ instead of v.

DEFINITION 10.3: $M(r, k) = \text{argmax}_x\ \Pi(r, k, 0, x, 0)$.

DEFINITION 10.4: $S(r, k) = \text{argmax}_x\ \tilde{\Pi}(r, k, 1, x)$.

$M(r, k)$ is the monopoly output of the incumbent firm with a commitment of (r, k), and $S(r, k)$ is its Stackelberg equilibrium output if entry occurs. The following lemma provides a characterization of monopoly output in the presence of pre-commitment.[4]

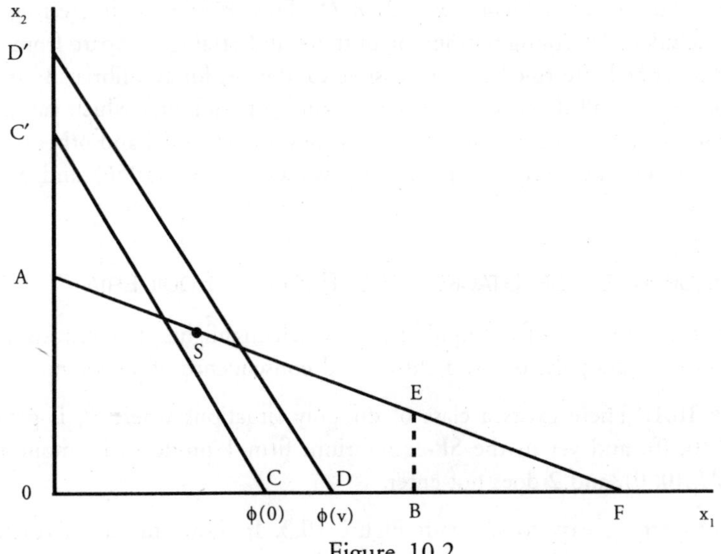

Figure 10.2

LEMMA 10.1: For all r in $[0, v]$ and all $k \geq 0$,

$$M(r, k) = \begin{cases} \phi(0), & \text{if } k < \phi(0) \\ \phi(r), & \text{if } k > \phi(r) \\ k, & \text{otherwise.} \end{cases}$$

REMARKS:

(i) $\phi(0)$ is, of course, the monopoly output or $M(0, 0)$.

(ii) Note that A10.3 (i) implies that $\phi(r)$ is strictly increasing in r.

(iii) By Lemma 10.1 and (ii), $M(r, k)$ is non-decreasing in k. Reasoning entirely analogous to Lemma 10.1 and the above remarks yield exactly the same characterization for $S(r, k)$. In particular $S(r, k)$ is non-decreasing in k. This is shown in proving the next result, which is the main one of this section.

THEOREM 10.1. *If* $(r_1^*, k_1^*, D^*, x_1^*, x_2^*)$ *is* an SP-equilibrium, then either (i) $x_1^* = M_1(r_1^*, k_1^*)$ *and* $D^* = 0$ (and, by implication, $x_2^* = 0$) *or* (ii) $x_1^* = S_1(0, 0)$ and $x_2^* = R(x_1^*)$.

Theorem 10.1 may be illustrated. In Figure 10.2, AEBF is the reaction curve of firm 2. CC′ and DD′ are 1's reaction curves with $r = 0$ and $r = v$ respectively. Lemma 10.1 implies that CD is the segment within which the monopoly equilibria of firm 1 with different commitment levels must lie. S is the usual Stackelberg point. Theorem 10.1 therefore asserts that the SP equilibrium must be either at S or on some point on CD. This theorem provides a partial characterization of SP-equilibria. Further properties of these equilibria are explored in the next section. Before going on to this, it is important to fully understand the significance of Theorem 10.1.

The intuitive argument behind Theorem 10.1 is clear: if entry is not to be deterred, then there is no point in making a costly commitment. Hence $x = S(0, 0)$. If entry is deterred, the firm is a monopolist, with the pre-commitment (r, k) necessary to deter entry. Hence, $x = M(r, k)$. This theorem is interesting for what it excludes. All configurations of outputs and strategies, apart from the ones just described, are ruled out as possible candidates for equilibria. By using this theorem we shall describe some special kinds of equilibria which can arise and which highlight the contrasts and similarities of our model and other works on entry barriers including the pioneering works of Bain (1956) and Sylos-Labini (1962).

10.4. PROPERTIES OF STACKELBERG PERFECT EQUILIBRIA

An interesting property of SP-equilibrium, which highlights its contrasts with other models of duopoly, is a straightforward consequence of Theorem 10.1.

PROPERTY 10.1: There exists a class of duopoly situations where B_1 is greater than M_1 (0, 0), and yet in the SP-equilibrium firm 1 produces its monopoly output, M_1 (0, 0), and 2 does not enter.

This property is easy to see with Figure 10.3. It shows the usual reaction functions of firms 1 and 2 and the iso-profit curve of firm 1 which passes

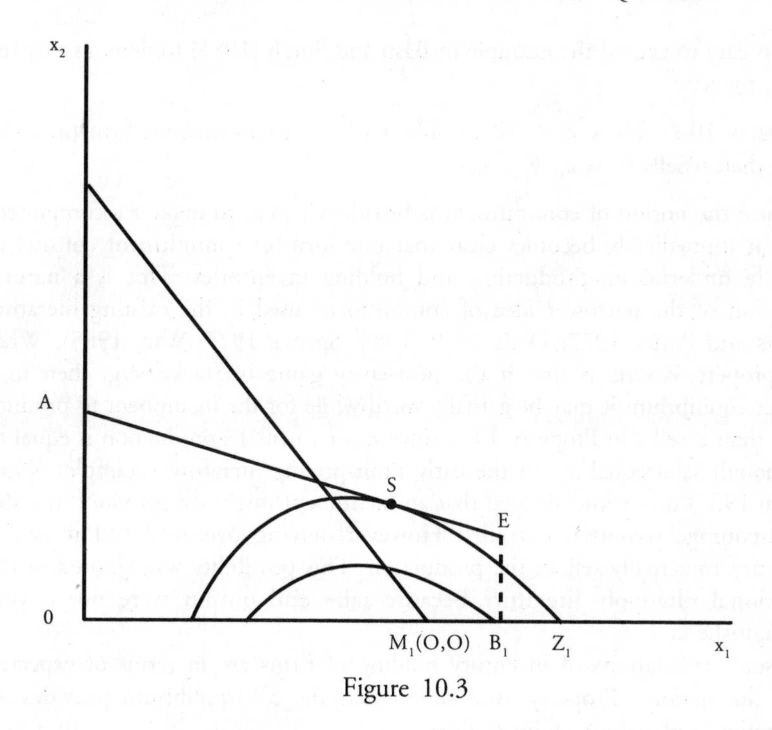

Figure 10.3

through the Stackelberg point S. Z_1 is the point where this iso-profit curve meets the x_1 axis.

If $B_1 < Z_1$, as is the case in Figure 10.3, 1 will produce its monopoly output M_1 (0, 0) and 2 will not enter. This is because if 2 enters, 1 is best off producing B_1, which means 2 will produce nothing. Knowing this, 2 will not enter. And knowing that 2 will not enter, 1 chooses its best output, M_1 (0, 0). This contrasts sharply with Spence (1977) and Dixit (1979) where, given a situation as in Figure 10.3, firm 1 would produce B_1. The models of Bain and Sylos-Labini also suggest that 1 would produce B_1.

In a perfect equilibrium, on the contrary, the fact that the incumbent possesses a strategy to costlessly eliminate a new firm, should it enter, is enough to guarantee that the new firm will not enter. The incumbent does not have to *actually adopt* the elimination strategy (which, in Figure 10.3, entails producing B_1).

The crucial role played by the existence of commencement costs, that is, fixed costs associated with positive production rather than entry (see discussion in Sections 10.1 and 10.2), is worth emphasizing here. If there were no such costs, then if firm 1 produced M_1 (0, 0), firm 2 would enter. This is because once entry occurs it has no further fixed costs and so the break in its reaction function at B_1 in Figure 10.3 is no longer there. Knowing this, firm 1 would accommodate the entrant by moving to the Stackelberg equilibrium output S_1(where S_1 is the projection of the point S on the horizontal axis). Since firm 2, in turn, knows this, it is in its interest to enter and settle for equilibrium at S.

It is easy to extend the example in Basu and Singh (1985) to demonstrate the next property.

PROPERTY 10.2: There exist SP equilibria where the incumbent firm produces more than it sells ($r_1 = v_1$, $k_1 > x_1$).

Once the notion of commitment is broadened so as to make r a component of it, it immediately becomes clear that one form of commitment consists of actually undertaking production and holding inventories. This is a natural extension of the narrower idea of commitment used in the existing literature (Caves and Porter 1977, Dixit 1979, 1980; Spence 1977; Ware 1985). What this property asserts is that if the post-entry game is Stackelberg, then in a perfect equilibrium it may be actually worthwhile for the incumbent to produce more than it sells. In Property 10.2, since $r_1 = v_1$, firm 1's production is equal to k_1, though sales equal x_1. In the early limit-pricing literature (example, Sylos-Labini 1962) it is acknowledged that an incumbent might overproduce in order to discourage potential entrants. However, having overproduced it is not necessary to actually sell all the production. This possibility was ignored in the traditional oligopoly literature because sales and output were not always distinguished.

Usual explanations of inventory holding of firms are in terms of expected price fluctuations. Property 10.2 shows that the SP-equilibrium provides an alternative explanation of inventories.

PROPERTY 10.3: There exist SP equilibria where the incumbent firm chooses a level of readiness greater than zero, but leaves no unutilized commitment ($0 < r_1 < v_1$, $k_1 = x_1$).

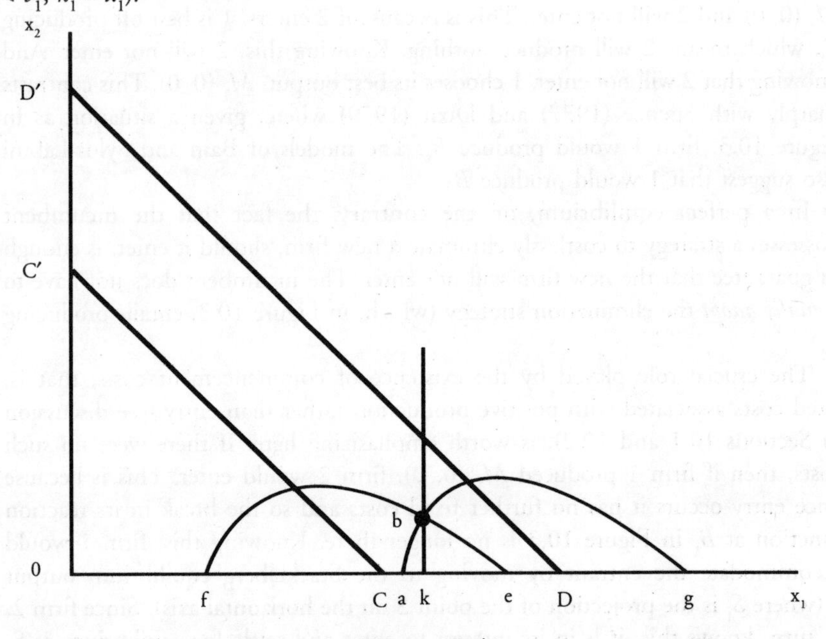

Figure 10.4

In this case the firm makes a commitment but the commitment provides no 'burden' because it is not left unutilized. At first sight the purpose of such commitment may seem unclear. But of course it is merely to ensure that cutting back production is costly to itself and thereby give a signal to the entrant that it will not be accommodating and cut back production should entry occur.

PROPERTY 10.4: There exist SP-equilibria in which $B_1 > Z_1$, yet firm 2 does not enter.

In Dixit (1979), as well as in Bain (1956), if $B_1 > Z_1$ then entry is ineffectively impeded and a usual Stackelberg equilibria is a *necessary* outcome. Property 10.4 suggests that in an SP-equilibrium this need not be so.

Properties 10.3 and 10.4 may be demonstrated geometrically. Figure 10.4 shows how to work out the iso-profit curves of a firm with commitment $(\hat{r}, \hat{k})$. Let $C'C$ and $D'D$ be firm 1's reaction with a commitment of 0 and $\hat{r}$ respectively. Let $\hat{k}$ be as shown in the diagram. Suppose *fbe* is an iso-profit curve corresponding to $C'C$ (i.e., with no commitment). Draw an iso-profit curve corresponding to $D'D$ (i.e., with a commitment of $r = \tilde{r}$ and very large k, larger than g) passing through b: This is shown by *abg* in the figure. Then for a firm with commitment $(\hat{r}, \hat{k})$, an iso-profit curve is shown by *abe.* Other iso-profit curves may be similarly constructed.

Now we may demonstrate Properties 10.3 and 10.4. Suppose in Figure 10.5, 1's iso-profit curve through B_1 intersects 2's reaction function at a point

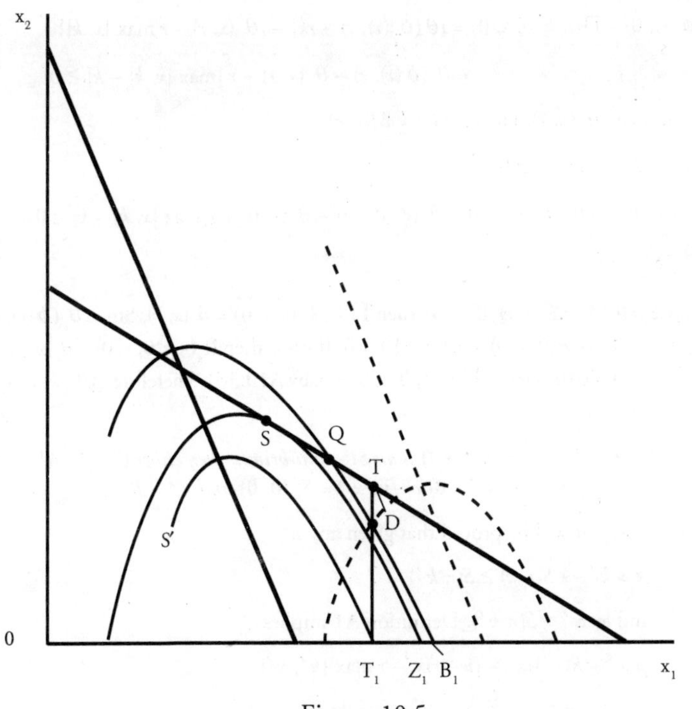

Figure 10.5

Q, to the left of Z_1, which is the point where the iso-profit curve through the Stackelberg point S touches the x_1-axis. Choose a point in between Z_1 and Q on the x_1-axis and call it T_1. For simplicity, assume 1's reaction function with $r_1 = v_1$ lies to the right of T. Such a reaction function and its associated iso-profit curves are shown by the broken lines. Suppose 1 sets $k_1 = T_1$ and $r_1 = v_1$. Then its iso-profit curves look like ADB_1, by the argument sketched above. Now if 2 enters, it will be best for 1 to produce B_1 (and be an iso-profit curve ADB_1). Knowing this, 2 will not enter. Hence 1 can act as a monopolist once it has committed (v_1, T_1). By Lemma 10.1, his sales will be T_1. Therefore, 1 will be on a higher iso-profit curve than at S (namely on curve T_1S'). This shows that there exists a strategy of firm 1 which dominates point S. Hence by Theorem 10.1, firm 2 will not enter and firm 1 will operate at M_1 (r_1, k_1) for some pre-commitment (r_1, k_1), thereby establishing both Property 10.3 and Property 10.4.

APPENDIX 10.1

LEMMA 10.1: For all r in $[0, v]$ and all $k \geq 0$,

$$M(r, k) = \begin{cases} \phi(0), & \text{if } k < \phi(0) \\ \phi(r), & \text{if } k > \phi(r) \\ k, & \text{otherwise.} \end{cases}$$

PROOF: Suppose $k > \phi(r)$. For any $x > 0$,

$$\Pi(r, k, 0, \phi(r), 0) - \Pi(r, k, 0, x, 0) = [\theta(\phi(r), r) - rk] - [\theta(x, r) - r \max\{x, k\}]$$

$$= \theta(\phi(r), r) - \theta(x, r) + r[\max\{x, k\} - k] \geq 0,$$

since $\phi(r) = \text{argmax}_x\, \theta(x, r)$. Hence $\phi(r) = M(r, k)$.

Suppose $k < \phi(0)$. For any $x \geq 0$,

$$\Pi(r, k, 0, \phi(0), 0) - \Pi(r, k, 0, x, 0) = \theta(\phi(0), 0) - \theta(x, 0) + r[\max\{x, k\} - x] \geq 0.$$

Hence $\phi(0) = M(r, k)$.

Finally, suppose $\phi(0) \leq k \leq \phi(r)$. If $x < k$, then $\Pi_x(r, k, 0, x, 0) = \theta_x(x, r)$. Since $\theta_x(\phi(r), r) = 0$ and $x < \phi(r)$, hence $\Pi_x(r, k, 0, x, 0) > 0$, by A 10.3(i). If $x > k$, then $\Pi_x(r, k, 0, x, 0) = \theta_x(x, 0)$. Since $\theta_x(\phi(0), 0) = 0$ and $x > \phi(0)$, hence $\Pi_x(r, k, 0, x, 0) < 0$, by A10.3(i). Therefore, $M(r, k) = \text{argmax}_x\, \Pi(r, k, 0, x, 0) = k$.

Q.E.D.

THEOREM A10.1: *If* $(r_1^*, k_1^*, D^*, x_1^*, x_2^*)$ *is an SP-equilibrium, then either* (i) $x_1^* = M_1(r_1^*, k_1^*)$ *and* $D^* = 0$ (*and, by implication,* $x_2^* = 0$) *or* (ii) $x_1^* = S_1(0, 0)$ *and* $x_2^* = R(x_1^*)$.

PROOF: As a first step, it will be proved that given any r,

(A1) $$k > k' \rightarrow S(r, k) \geq S(r, k').$$

Suppose $k > k'$ and let $x' \equiv S(r, k')$. Definition A1 implies

(A2) $$p(x' + R(x'))x' - (v - r)x' - r \max\{k', x'\}$$

$$> p(\hat{x} + R(\hat{x}))\hat{x} - (v - r)\hat{x} - r \max\{k', \hat{x}\},\ \forall\, \hat{x} < x'$$

The strict inequality in (A2) is because $S(r, k')$ is unique by virtue of Assumption A10.3(ii). It is easily checked that if k' is replaced by k, inequality (A2) remains unchanged. Thus, even with k, firm 1 finds that x' earns a greater profit than all $\hat{x} < x'$. Hence as k' is replaced by k, firm 1 will not choose a smaller output, thereby establishing (4).

Let Q be the set of commitments of firm 1, given which, 2 prefers to stay out of the industry:

$$Q = \{(r, k) \mid S(r, k) \geq B_1\}.$$

First, consider the case where $(r^*, k^*) \in Q$. If $D = 1$, $x_2 = 0$ and firm 2 earns no profit. If $D = 0$, $x_1 = M(r^*, k^*)$ and 2 earns zero profit. By A10.2 (ii), firm 2 chooses $D^* = 0$, and $x_1^* = M(r^*, k^*)$.

Next suppose $(r^*, k^*) \notin Q$. If $D = 1$, then $x_1 = S(r^*, k^*) < B_1$. Firm 2 earns positive profit. Hence $D^* = 1$, $x_1^* = S(r^*, k^*)$ and $x_2^* = R(x_1^*)$. Furthermore, $\tilde{\Pi}$ is decreasing in k for $k > x$. This, together with (A1)—which implies that (r^*, k) $\notin Q$ for $k < \tilde{k}^*$—implies $k^* \leq x_1^*$. But then, $\tilde{\Pi}(r^*, k^*, 1, x_1^*) = p(x_1^* + R(x_1^*))$ $x_1^* - v_1 x_1^* = \tilde{\Pi}(0, 0, 1, x_1^*)$. Suppose now that for some x_1,

(A3) $\quad \tilde{\Pi}(0, 0, 1, x_1) > \tilde{\Pi}(0, 0, 1, x_1^*).$

Since $(0, 0) \notin Q$, (A3) implies that firm 1 can do better by committing (0, 0), that is, nothing. Hence, (r^*, k^*) cannot be part of a perfect equilibrium outcome. This contradiction establishes (A3) as false. Thus $x^* = \operatorname{argmax}_x \tilde{\Pi}(0, 0, 1, x) \equiv S_1(0, 0)$.

Q.E.D.

Notes

1. A different direction to pursue would be to consider different *kinds* of capital and their role in entry-deterrence. See Eaton and Lipsey (1981) for an interesting analysis along these lines.
2. The case of inventory holding can be explained and interpreted in another way: It could be thought of as the case of committing the most extreme product-specific capital, namely, the product itself (see Ware 1985).
3. We are grateful to Avinash Dixit for emphasizing to us the role of these two different kinds of fixed costs.
4. Proofs of Lemma 10.1 and Theorem 10.1 are available in Appendix 10.1.

References

Bain, J.S. 1956. *Barriers to New Competition*. Cambridge: Harvard University Press.

Basu, K. and N. Singh. 1985. 'Commitment and Entry-Deterrence in a Model of Duopoly'. *Economics Letters* 18.

Bulow, J., J. Geanakoplos, and P. Klemperer. 1985. 'Holding Idle Capacity to Deter Entry', *Economic Journal* 95.

Caves, R. and M. Porter. 1977. From Entry Barriers to Mobility Barriers'. *Quarterly Journal of Economics* 91.

Dixit, A. 1979. 'A Model of Duopoly Suggesting a Theory of Entry Barriers'. *Bell Journal of Economics* 10.

——. 1980. 'The Role of Investment in Entry-Deterrence'. *Economic Journal* 90.

Eaton, B.C. and R. Lipsey. 1981. 'Capital, Commitment and Entry Equilibrium'. *Bell Journal of Economics* 12.

Friedman, J. 1977. *Oligopoly and the Theory of Games*. Amsterdam: North Holland.

Saloner, G. 1985. 'Excess Capacity as a Policing Device'. *Economics Letters* 18.

Salop, S.C. 1979. 'Strategic Entry Deterrence'. *American Economic Review*. Papers and Proceedings 69.

Spence, A.M. 1977. 'Entry, Capacity, Investment and Oligopolistic Pricing'. *Bell Journal of Economics* 8.

Sylos-Labini, P. 1962. *Oligopoly and Technical Papers*. Cambridge: Harvard University Press.

Ware, R. 1985. 'Inventory Holding as a Strategic Weapon to Deter Entry'. *Economica* 52.

11 Collusion in Finitely Repeated Oligopolies

11.1. Introduction

The explanation of collusion in finitely repeated oligopolies is problematic because the backward-induction argument predicts the Cournot equilibrium in each period. However, it is intuitively troublesome to think of rational firms, year after year, making a lower profit because of this.[1]

Economists have responded to this conflict between intuition and reason by constructing a whole range of models explaining collusion. One approach is to consider an infinitely repeated oligopoly and *treat it* as an approximation for very long, finite repetitions (see Friedman 1971; Green and Porter 1984; Abreu et al. 1986; Rotemberg and Saloner 1986). The second avenue is to retain the *finite*-repetition assumption as realistic but modify the assumption of individual rationality (see Radner 1980, and in a more abstract context, Kreps et al. 1982).[2]

There is a third line of analysis which seems to have been ignored in the large literature on collusion. This involves giving more structure to the Cournot *stage* game. Once we allow for this, collusive behaviour becomes explicable without having to sacrifice the assumption of *finite* length of interaction and full rationality (in fact, its common knowledge) among the firms. The aim of this chapter is to explore this line of explanation.

Production takes time and it seems reasonable to suppose that an oligopolist can observe whether its competitors are planning to produce a lot or a little; and can respond to this by adjusting its own production plans before the product is

From *International Journal of Industrial Organization* 10, 1992: 595–609.

I have benefited from discussions with Geir Asheim, Abhijit Banerji, David Cooper, Avinash Dixit, Birgit Grodal, Christian Schultz and Jorgen Weibull and from a presentation at the Sixth World Congress of the Econometric Society in Barcelona.

finally made available on the market. I shall in the present chapter allow for this kind of structure in a somewhat stylized manner, which is a variant of Saloner (1987). This entails thinking of each stage game as broken up into two substages. In the first substage each firm produces some amount. They observe this and in the second substage produce more or dispose of any amount of the output produced in the first substage. Then their total production is offered on the market and price and profits are determined in the usual Cournot style. I shall refer to this as the *modified* Cournot game. Note that the modified Cournot game is itself an extensive-form game. The model in Section 11.2 which combines Saloner (1987) and Friedman (1985) is an illustration of this.

If the modified Cournot game is played a finite number of times, collusion becomes possible under subgame perfection. We have to however, be careful in defining 'subgame perfection' since the full game consists of repetitions of an *extensive*-form game. Hence, we have to consider histories of not just n stage games but also n stage games and a 'half' of the $(n+1)$th stage game.

My model can be modified in many ways: by using different trigger strategies, such as in Benoit and Krishna (1985), and other stage games, such as in Basu (1990). Recently robustness questions have been raised about Saloner's model (see Pal 1991). Section 11.3 argues that those features of Saloner's model which are used here are relatively robust. The most untenable feature of Saloner's model is the assumption that it is impossible to dispose of goods once produced. Section 11.4 relaxes this assumption. Though the case I discuss is the polar one where destroying goods is costless, the *method* can be used to analyse more realistic intermediate cases.

Though the focus of this chapter is on *collusion* in oligopolistic industries, a subject of considerable interest in the industrial organization literature (see Green and Porter 1984; Abreu et al. 1986, one can raise more general questions concerning the *set* of outcomes that are supportable in equilibrium. Though subgame perfection in my model places more severe restrictions on strategies than in models where a *one*-shot game is played repeatedly, a Benoit–Krishna type argument can be constructed even here. Hence, any pay-off which is feasible and individually rational will, arguably, be supportable in equilibrium if the number of repetitions is large enough. This last qualification is extremely important and is the reason why Benoit and Krishna call their result a '*limit* perfect folk theorem'. While such a result is of game-theoretic interest, its appeal in the industrial-organization framework considered here is limited since it hinges on our being able to repeat the stage game arbitrarily many times, which violates the 'spirit' of finiteness.

11.2. The Model and Collusive Equilibria

11.2.1. Basic Concepts and Definitions

The aggregate inverse demand function facing the industry (in each period) is given by $p = p(x)$ and firm i's cost function is $c_i(x_i) = dx_i$, $i = 1,2$. We shall assume throughout that the cost function is linear. If the total production of the firms is x_1 and x_2, then firm i's profit is given by

(11.1) $$\pi_i(x_1, x_2) \equiv x_i p(x_1 + x_2) - d_i x_i,\ i = 1,2.$$

The *reaction function* of i is defined as

(11.2) $$\phi_i(x_j) \equiv \underset{x_i}{\operatorname{argmax}}\ \pi_i(x_1, x_2),\ i \neq j,\ i = 1,2.$$

It is being assumed that for all x_j, $\phi_i(x_j)$ is unique.

The output pair, $x^N = (x_1^N, x_2^N)$ will be called the *Cournot Outcome*[3] if and only if $\pi_1(x^N) \geqq \pi_1(x_1, x_2^N)$, for all x_1 and $\pi_2(x^N) \geqq \pi_2(x_1^N, x_2)$, for all x_2. We shall throughout use x^N to denote the Cournot outcome. We define the Cournot profit as follows:

$$\pi_i^N \equiv \pi_i(x^N).$$

The *Stackelberg outcome with 1 as leader* is $\bar{x} = (\bar{x}_1, \bar{x}_2)$ if $\bar{x}_1 = \operatorname{argmax}_{x_1} \pi_1(x_1, \phi_2(x_1))$ and $\bar{x}_2 = \phi_2(\bar{x}_1)$. The Stackelberg outcome with 2 as leader is defined symmetrically. The profit of i when j is Stackelberg leader is denoted by π_{ij} and this is defined as follows:

$$\pi_{ij} \equiv \pi_i(\bar{x}),$$

where $\bar{x}$ is the Stackelberg outcome with j as leader.

Figure 11.1 illustrates the above functions and outcomes, with N denoting the Cournot outcome and S^i the Stackelberg outcome with i as leader. The curves marked π_i^N, $i = 1, 2$, are iso-profit curves through N.

Throughout the analysis I shall assume that the demand function is sufficiently well-behaved for the following to be true.

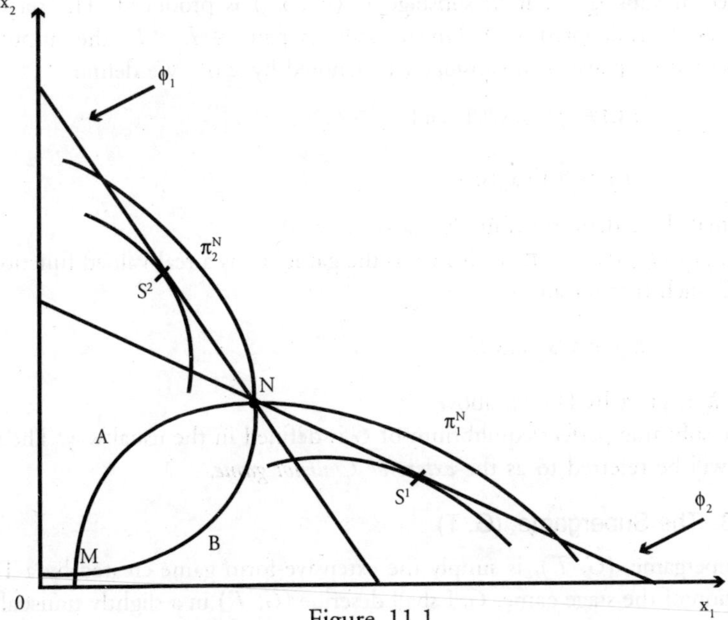

Figure 11.1

(i) For each i, π_i is continuous and strictly quasi-concave with respect to x_i.
(ii) $\pi_1(x_1, \phi_2(x_1))$ is strictly quasi-concave with respect to x_1 and $\pi_2(\phi_1(x_2), x_2)$ is strictly quasi-concave with respect to x_2.
(iii) $\partial\phi_i/\partial x_j > -1$, $i \neq j$, $i = 1,2$.

Condition (iii) implies that reaction functions are 'stable' in the textbook sense and the Nash outcome is unique. Condition (ii) has the advantage of ensuring that there is a unique Stackelberg equilibrium and that the Stackelberg leader's profit increases monotonically as we move along the follower's reaction curve towards the Stackelberg equilibrium. This condition is used in Basu and Singh (1990).

11.2.2. The Stage Game, G

As discussed above, we want to give some temporal structure to the usual Cournot game by allowing for the fact that production takes time and firms can spy on each other's production effort and make suitable adjustments. This is captured formally by using Saloner's (1987) formulation. Assume that each period or stage is broken up into two substages. In substage 1, the firms produce $(x_1^{\,1}, x^1_{\,2})$. Then, having observed this, in substage 2, they produce $(x_1^{\,2}, x^2_{\,2})$. We assume, $x_i^t \geqq 0$, for all t, for all i. This implies that in the second substage, firms can only add to their output. This is a temporary assumption which allows me to use Saloner's theorem. It is possible to allow firms the freedom to destroy output in the second substage without losing my central claim concerning collusion. This is shown in Section 11.4.

If s_i is a *strategy* of firm i, then s_i specifies a production level, $x_t^{\,1}$, in substage 1 and a function on all output pairs such that $s_i(x^1_{\,1}, x^1_{\,2})$ is the output i would produce in substage 2, if in substage 1, $(x^1_{\,1}, x^1_{\,2})$ is produced. The set of all strategies of i is denoted by E_i. Given a strategy pair $s \in E_1 \times E_2$, the output that is produced by player i in substage t is denoted by $x_i^t(s)$. We define

$$(11.3) \qquad x^t(s) \equiv (x_1^{\,t}(s), x_2^{\,t}(s)), \text{ and}$$

$$x_i(s) \equiv x_i^{\,1}(s) + x_i^{\,2}(s).$$

Note that these definitions imply $s_i(x^1(s)) = x^2_{\,i}(s)$.

The *pay-off function,* $\tilde{\pi}_i$, of firm i in the game, G, is a real-valued function on $E_1 \times E_2$ such that for all $s \in E_1 \times E_2$,

$$\tilde{\pi}_i(s) \equiv \pi_i(x_1(s), x_2(s)),$$

where π_i is giver by (11.1), above.

The subgame-perfect equilibrium of G is defined in the usual way. The stage game will be referred to as the *extensive Cournot game.*

11.2.3. The Supergame, (G, T)

The supergame, (G, T), is simply the extensive-form game created by a T-fold repetition of the stage game, G. I shall describe (G, T) in a slightly unusual way which makes no explicit reference to the stage game. This is merely a matter of convenience.

Note that (G, T) involves $2T$ substages which we shall refer to as substages $1,2,\dots,2T$. Every consecutive pair of substages, $(t, t+1)$, where t is odd, constitutes a stage.

Let x_i^t be firm i's output in substage t. We write $x^t \equiv (x_1^t, x_2^t)$. A *t-substage history* is a sequence, $(x^1, \dots, x^t)$, of outputs in all substages up to the tth one. I shall at times refer to a t-substage history as $h(t)$. Let $H(t)$, $t > 1$, be the set of all t-substage histories. Define $H(0) = \{\phi\}$ and $H = \cup_{t=0}^{2T-1} H(t)$.

A strategy, f_i, of firm i in the supergame, (G, T), is a mapping, f_i: $H \rightarrow R_+$, where R_+ is the set of non-negative real numbers. $f_i(h(t))$ denotes the output i will produce in substage $t+1$, given that the history thus far is described by $h(t)$; and $f_i(\phi)$ is i's output in substage 1, that is, at the start of the supergame.

Let F_i be the set of all strategies of i in (G, T). If $f \in F_1 \times F_2$, then $\sigma(f)$ denotes the path of outputs that would be produced if f is the strategy pair being played; $\sigma(f)$ is defined in the usual way.

Finally, we have to define the *pay-off function*, P_i, of player i. Let D be the set of odd numbers in $\{1,2,\dots,2T\}$. If $f \in F_1 \times F_2$, and $\sigma(f) = (x^1,\dots,x^{2T})$, then

$$P_i(f) \equiv \sum_{t \in D} \pi_i(x_1^t + x_1^{t+1}, + x_2^t + x_2^{t+1}).$$

The use of the stage game is implicit in the specification of P_i.

The *average pay-off* earned by i in (G, T) while playing f is denoted by $p_i^T(f)$ and defined as $P_i^T(f) \equiv P_i(f)/T$.

The equilibrium criterion that will be used is subgame perfection. To define this formally we need some more notation. Given any strategy pair, f, and a history, $h(t)$, the path generated in the $2T - t$ remaining substages is defined in the usual way and denoted by $\sigma(f, h(t))$. I shall use P_i to define pay-offs in subgames. If f is the strategy pair being used and $h(t)$ is the history after which the $(2T - t)$-substage subgame is being considered, then $\bar{P}_i(f, h(t))$ is i's pay-off in the subgame. This is defined as follows.

$$\bar{P}_i(f, h(t)) \equiv \sum_{t \in D} \pi_i(x_1^t + x_1^{t+1}, + x_2^t + x_2^{t+1}),$$

where $h(t) = (x^1,\dots, x^t)$ and $\sigma(f, h(t)) = (x^{t+1},\dots, x^{2T})$.

A strategy pair f^* is a *subgame-perfect equilibrium* of (G, T), if and only if, for all $h(t) \in H$, for all $i \in \{1,2\}$, and for all $f_i \in F_i$,

$$\bar{P}_i(f^*, h(t)) \geqq \bar{P}_i(f^*/f_i, h(t)),$$

where f^*/f_i is the strategy pair formed by replacing the ith element of f^* *with* f_i.

11.2.4, The Theorem

Let $(x^*, x_2^*) = x^*$ be such that $\pi_i(x^*) \geqq \pi_i^N$, $i = 1,2$. In other words, x^* is any point in the area enclosed by MANBM in Figure 11.1.

THEOREM 11.1. (*a*) There exists T such that there is a subgame perfect equilibrium, f^*, in (G, T) which results in x^*_i being produced by i in the first stage game. That is, the outputs produced by i in substages 1 and 2 add up to x_i^*.

(*b*) For all $\varepsilon > 0$, there exists T such that f^* is a subgame perfect equilibrium of (G, T) and $p^T_i(f^*) > \pi_i(x^*_1, x^*_2) - \varepsilon$.

Part (a) of the theorem asserts that in the early stage games collusive behaviour can be supported under subgame perfection. Part (b) asserts that as the supergame is made longer and longer the stages with collusive behaviour can be so much more numerous than the stages with non-collusive outcomes that the average (that is, per-stage) pay-off converges to the pay-off under collusion.[4]

Since the joint-profit-maximizing locus in the usual Cournot game (the contract curve) passes through the area MANMB, the theorem suggests that joint-profit-maximizing outcomes are possible (in the limit) under full, individual rationality and finite repetitions of the modified Cournot game.

11.2.5. The Proof

A critical input in the proof of Theorem 11.1 is a proposition of Saloner's (1987), which is restated here as Lemma 11.1. This is concerned with outcomes in the stage game, G, described above. Let $(x_1(i), x_2(i))$ be the Stackelberg outcome with i as leader, and, as before, let (x_1^N, x_2^N) be the Nash outcome.

Lemma 11.1. In the stage game, G, each of the following three production paths is the outcome of a strategy pair, which is subgame perfect:

(*a*) In the first substage, the firms produce $(x_1(1), x_2(1))$ and in the second substage they produce (0,0).
(*b*) In the first substage, the firms produce $(x_1(2), x_2(2))$ *and in the second substage they produce (0,0).*
(*c*) In the first substage, the firms produce $[x_1^N, x_2^N]$and in the second substage they produce (0,0).

This lemma asserts that the Cournot outcome as well as the two Stackelberg outcomes can be supported under subgame perfection if the Cournot game is played in two stages as in *G*. For a detailed proof the reader is referred to Saloner (1987). I shall give here a sketch of the argument which will help prepare the ground for a generalization of Saloner's result to the case with free disposal of output.

To prove Lemma 11.1 suppose $(x_1^1, x_2^1) = x^1$ has been produced in substage 1. What will happen in substage 2 depends on which of the following four regions of Figure 11.2, x^1 happens to be in: region $ON_1\,NN_2$; region $N_1\,NA_1$; region $N_2\,NA_2$; region north-east of A_2NA_1. Under subgame perfection, if x^1 is in region ON_1NN_2, substage 2's output must be such that total output occurs at N; if x^1 is in region $N_1NA_1(N_2NA_2)$ substage 2's output must be such that total output occurs on firm 2's (firm 1's) reaction function vertically above (horizontally

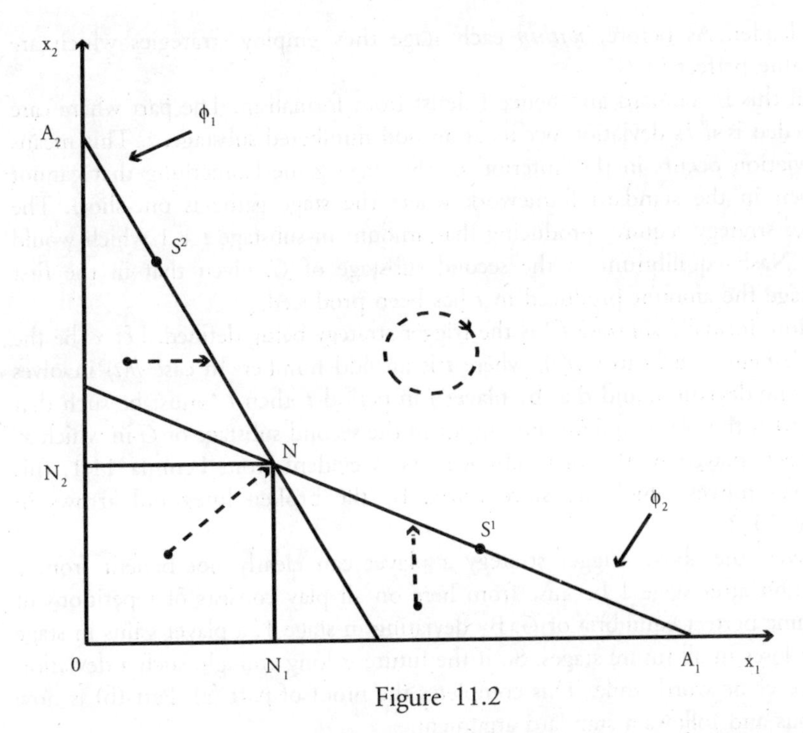

Figure 11.2

to the right of) x^1. If x^1 is in the north-east of A_2NA_1, then total output remains at the same place. The dashed lines with arrows summarize this in Figure 11.2.

The proof of this is simple, and requires keeping in mind that subgame perfection implies Nash-equilibrium play in the second substage and if x^1 has been produced in substage 1, then only the region north-east of x^1 is feasible in substage 2.

Now turn to substage 1. Suppose firms 1 and 2 plan to produce S^1 in substage 1. If 1 deviates, that is, the production in substage 1 is a horizontal displacement from S^1, then, given what we already know will happen in substage 2, the final outcome will occur either somewhere on the line NS^1 or somewhere to the right of S^1. In either case, firm 1 cannot do better through a deviation. It is easy to check that neither can 2 do better. Hence, S^1 is an outcome of a subgame perfect equilibrium. The proofs for S^2 and N are similar. This completes the proof of Lemma 11.1.[5]

The proof of Theorem 11.1, which will be stated only in outline, entails the construction of a suitable discriminating trigger strategy. Let x^* be as in Theorem 11.1. Consider the following 'path': in the first stage game x^* is produced—this is achieved by producing x^* in substage 1 and 0 in substage 2. Then on they produce x^N in each stage—the substage-wise break-up within each stage is achieved in a manner which is subgame perfect in G. The trigger strategy consists of proceeding on the above path as long as no deviation by a single player occurs at any substage. If i deviates alone in some substage, then in all remaining stages they play so as to achieve the Stackelberg outcome with 3 –

i as leader. As before, *within* each stage they employ strategies which are subgame perfect in G.

All this is standard and hence I desist from formalism. The part where care is needed is if i's deviation occurs at an odd numbered substage, t. This means a deviation occurs in the 'interior' of the stage game (something that cannot happen in the standard framework where the stage game is one-shot). The trigger strategy requires producing that amount in substage $t + 1$ which would be a Nash equilibrium in the second substage of G, given that in the first substage the amount produced in t has been produced.

More formally suppose f^* is the trigger strategy being defined. Let x^t be the last element of a history $h(t)$, where t is an odd number. In case $h(t)$ involves only one deviation, and that by player j in period t, then f^* must be such that $f^*(h(t))$ is the Nash equilibrium output in the second substage of G in which x^t has been played in the first substage. As is evident from Lemma 11.1, this involves moves which are summarized by the broken lines and arrows in Figure 11.2.

Given the above trigger strategy a player can clearly not benefit from a deviation after stage 1 because from here on all play consists of repetitions of subgame perfect equilibria of G. By deviating in stage 1, a player gains in stage 1 but loses in all future stages. So if the future is long enough, such a deviation can never be worthwhile. This completes the proof of part (a). Part (b) is now obvious and follows a standard argument.

11.3. Extensions and Robustness

As explained earlier, the model in Section 11.2 is an *illustration* of a more general idea. The model can be extended and modified in several directions. A generalization to the case of oligopoly with n firms is fairly easy and will not be pursued here.

One extension—which amounts to a robustness check on the Saloner model—is to allow for differences in the costs of production in the two substages of G. Return to the description of G and suppose the costs of producing one unit in substages 1 and 2 are, respectively, c_1 and c_2. A recent paper by Pal (1991)[6] shows that Saloner's theorem depends critically on his assumption that $c_1 = c_2$. Consider first $c_1 < c_2$. It is easy to check that in this case the only subgame-perfect equilibrium of G is the Cournot outcome. Hence in such a case my collusion argument would not be possible to construct.

What happens in the other case, that is, $c_1 > c_2$? Pal shows that if the gap is not 'too large', then the only subgame-perfect equilibria of G are the two Stackelberg outcomes. While this does damage the 'continuum claim' in Saloner (1987), *it retains the feature of Saloner's model which is needed for my model of Section* 11.2. While it is true that the simple trigger strategy of Friedman (1985) cannot be used to sustain collusive behaviour, we can use more sophisticated but standard trigger strategies which require rotating between the two Stackelberg outcomes in the last games. And note that if c_1 is just a little higher than c_2, there will exist open neighbourhoods around c_1 and c_2 such that as long as the costs are anywhere in these neighbourhoods the two Stackelberg outcomes will be supportable by subgame-perfect-equilibrium strategies in the extensive Cournot games.

There is a more general robustness issue concerning Saloner's model. Let us return to the game G with $c_1 = c_2$. In this section I shall call a strategy pair a *sturdy equilibrium* if it is subgame perfect and neither of these strategies is weakly dominated. Clearly, if a strategy pair is not a sturdy equilibrium, it cannot be a *trembling-hand* perfect equilibrium. It is possible to show that only the two Stackelberg outcomes can be supported by *sturdy equilibrium* strategies. In other words, all the other Saloner equilibria entail the use of a weakly dominated strategy by at least one player. Fortunately, as the remarks in the above paragraph indicate, this does not change the claims of the present chapter concerning the possibility of collusion.

11.4. Free Disposal

All this time it was being assumed that goods, once produced, cannot be destroyed. This is of course an unrealistic assumption. The aim of this section is to demonstrate that the claims of this chapter remain intact even if we allow for free disposal. Consider the description of the game G in Section 11.2.2. We will here relax the assumption that $x_i^2 \geqq 0$ for all i. Instead, it will be assumed that $x_i^2 \in [-x_i^1, \infty)$. That is, in the second substage a firm can destroy any amount of the production undertaken in the first substage or produce more. In addition, destruction of output is costless, that is, we have free disposal. I shall call this game G^*.

Before going into the analytics, note that G^* is best viewed as a polar case of a more realistic model in which disposal is not completely costless. There are products which, once produced, are difficult to destroy and costly to store. Hence, in reality, whereas production usually entails substantial costs, even destruction involves some costs, though generally less than what is incurred for production. Hence, the alternative assumptions of destruction being impossible and it being costless are both polar cases. We are about to analyse what seems to me to be the one closer to reality, namely, the latter case. More importantly, once my *method* of analysis is understood, it will be evident that the non-polar case can be analysed by the same method. The technique consists of using the real and the 'pseudo-reaction function' (defined below), in a manner originally used by Dixit (1980) (see also Basu and Singh 1990). As will be clear later, the non-polar case is basically one where the relevant pseudo-reaction function moves further away from the real one.

Turning now to the analysis of G^*, I shall briefly sketch a proof which demonstrates that even after we allow for free disposal, the stage game has at least three subgame perfect equilibria: (i) the usual Cournot outcome, (ii) an equilibrium where firm 1 earns a smaller profit than in the Cournot outcome, and (iii) one where 2 earns a smaller profit than in the Cournot outcome. Once

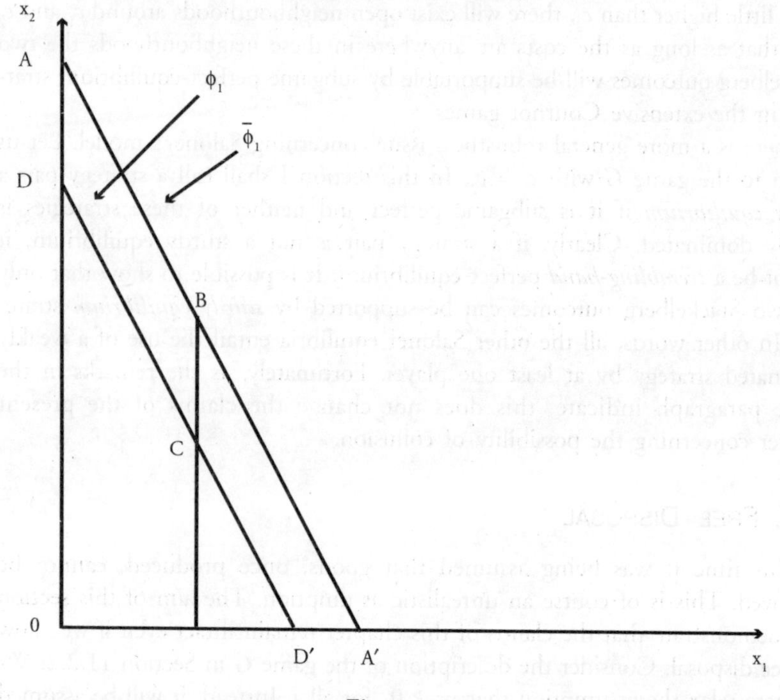

Figure 11.3

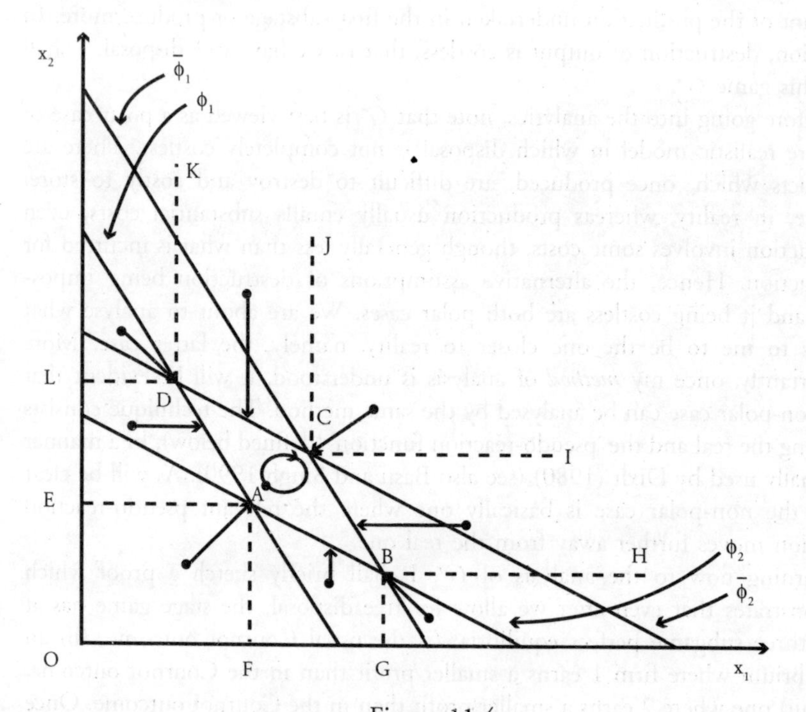

Figure 11.4

this is proved, it is obvious that a trigger strategy can be constructed to support collusive behaviour under subgame perfection.

To sketch a proof of the claim made in the above paragraph, let me introduce some new terminology. By a pseudo-reaction function of firm i, I shall here mean a reaction function of i under the assumption that i's cost of production is 0. If $(\bar{\phi})_i(\cdot)$ is i's pseudo-reaction function then

$$\bar{\phi}_i(x_j) = \underset{x_i}{\operatorname{argmax}}\, x_i p(x_1 + x_2).$$

Contrast this to (11.2). In Figure 11.3, firm 1's reaction and pseudo-reaction functions are both illustrated.

Suppose firm 1 produces $x_1^{\,1}$ in substage 1. Then, given free disposal, firm 1's reaction function in terms of total output produced at the end of substage 2 is given by the line ABCD′. This is because to produce more than $x_1^{\,1}$, the marginal cost is c_1 and DD′ is the relevant reaction function. To produce a total amount less than $x_1^{\,1}$, the marginal cost is zero. In other words, if firm 2 produces a total output of x_2, firm 1's optimal response in terms of the total output produced in the two substages, given that it has already produced $x_1^{\,1}$ in substage 1, can be read off the line ABCD′. Firm 2's aggregate reaction function is similarly worked out.

Now it is easy to work out for each substage 1 production, $(x_1, x_2) = x^1$, where the Nash equilibrium at the end of substage 2 will occur. Following the method used in Section 11.2.5, the result of what happens in substage 2 after every possible history is summed up in Figure 11.4. Essentially there are nine regions given by the areas enclosed in EAF, FABG, GBH, HBCI, ICJ, JCDK,

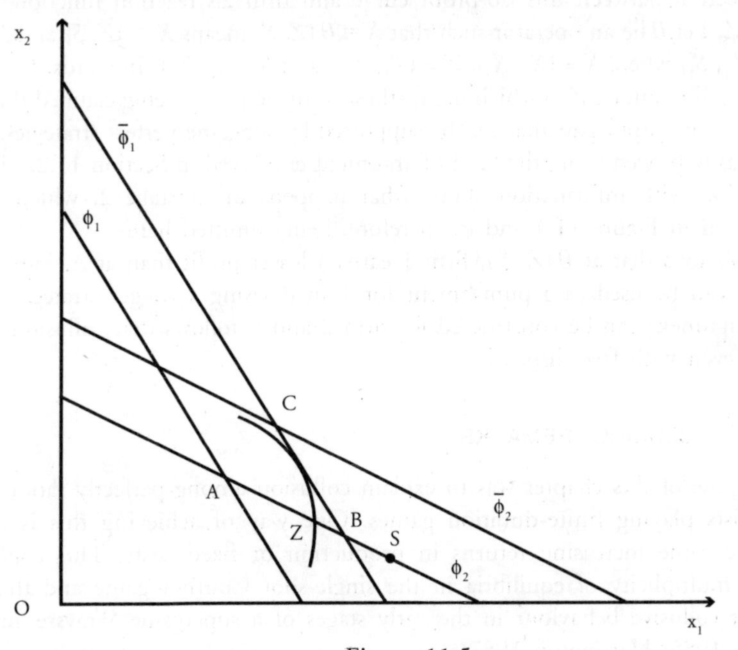

Figure 11.5

KDL, LDAE, and ABCD. Given that x^1 is in each of these regions, what happens in substage 2 is shown by the arrows. For instance, if x^1 is in ABCD, $x_1^2 = x_2^2 = 0$. These are easy to check using the aggregate-reaction function idea of Figure 11.3.

It should be immediately obvious that the subgame-perfect equilibria of the two-substage Cournot game can occur only on or within ABCD. This has an important implication. Suppose that the cost of production actually happened to be zero. That is, $c_i(x_i) = 0$, for all x_i. Then ϕ_i and $\bar{\phi}_i$ would coincide. Hence the area ABCD would collapse to a point, namely the usual Cournot-equilibrium point. It follows that the only subgame-perfect equilibrium in such a model would be the Cournot outcome.

I shall here continue with the realistic assumption of the marginal cost of production being positive. Using the second-substage behaviour, summed up in Figure 11.4, it is obvious that point A (the Cournot outcome) is a subgame-perfect equilibrium of G^*. I shall now isolate for each firm i, a subgame-perfect equilibrium of G^* where i earns less profit than at A. Without loss of generality, I focus on firm 1.

Continuing with the informal exposition, let us isolate three points of significance in Figure 11.5. Let B be the point where firm 1's pseudo-reaction function intersects 2's reaction function. Let S be the Stackelberg outcome with 1 as leader. Next let C be the point of intersection between the pseudo-reaction functions of 1 and 2. Pick the iso-profit curve of firm 2 which is tangential to the line segment BC (that is, the iso-profit curve which maximizes 2's profit and makes contact with 1 s pseudo-reaction function between B and C). The point of intersection between this iso-profit curve and firm 2's reaction function is labelled Z. Let θ be an operator such that $X \equiv \theta(Z, S)$ means $X \in \{Z, S\}$ and $X_1 = \min\{Z_1, S_1\}$ where $X = (X_1, X_2)$, $Z = (Z_1, Z_2)$, and $S = (S_1, S_2)$. In words, $\theta(Z, S)$ is that point among Z, S which lies furthest to the left. It is being claimed that $\theta(Z, S)$ is an output pair that can be supported by subgame-perfect strategies.[7] This is easily proved using the kind of argument employed in Section 11.2.5 in conjunction with information about what happens in substage 2, which is summarized in Figure 11.4 and is, therefore, being omitted here.

It is obvious that at $\theta(Z, S)$, firm 1 earns a lower profit than at A. Hence $\theta(Z, S)$ can be used as a punishment for 1 in devising a trigger strategy. A similar argument can be constructed for firm 2 and it follows that collusion is possible even with free disposal.

11.5. Concluding Remarks

The purpose of this chapter was to explain collusion among perfectly rational oligopolists playing finite-duration games. One way of achieving this is to introduce some increasing returns in production or fixed costs. This could create a multiplicity of equilibria in the single-shot Cournot game and thus allow for collusive behaviour in the early stages of a supergame (Fraysse and Moreaux 1985; Harrington 1987).

The present chapter, however, uses a stage game which is itself an *extensive-form* game. The supergame therefore arises out of repetitions of an extensive-form game. One has to be careful in defining subgame perfection in this context because we have to consider subgames which arise after *n and a half* stage games. Collusion is then shown to arise under subgame perfection.

There was another objective in the chapter. In the stage game considered here, the existing literature assumes that agents cannot dispose of goods once they are produced. This is clearly an unrealistic assumption. In Section 11.4, therefore, it was assumed that disposal is free and it was shown that though this results in a shrinkage of the set of equilibria, the shrinkage is not enough to rule out collusion.

NOTES

1. For a related problem concerning entry-deterrence Selten (1978) expresses a similar sentiment. So strong does he find the intuitive argument's challenge to backward induction that he describes this as a paradox and asserts that 'up to now I met nobody who said that he would behave according to the induction theory'.
2. This raises some troublesome methodological questions (see, for example, Binmore 1987; Basu 1990a), but my concern in this chapter is more mundane.
3. I purposely use the word 'outcome' instead of 'equilibrium', because the Cournot outcome is not what the present model predicts will happen necessarily. It is merely a definition which will come in handy later.
4. This means that the average profit earned by a firm will be between the joint-profit maximizing solution and the standard Cournot outcome, converging towards the former only as the time horizon becomes long. It is instructive to compare this with Slade's (1987) interesting empirical study based on competition among gasoline service stations in Vancouver. She finds that the stations earn less than the monopoly profit but more than in 'non-cooperative single-period solutions'.
5. Actually all points on the line segments NS^1 and NS^2 can be supported under equilibrium.
6. For some related research currently being done at Princeton, see Banerji and Cooper (1991).
7. In fact, all points on firm 2's reaction function between $\theta(Z, S)$ and A are supportable under subgame perfection. And, of course, a symmetric claim is true with the firms reversed.

REFERENCES

Abreu. D., D. Pearce, and E. Stacchetti. 1986. 'Optimal Cartel Equilibria with Imperfect Monitoring'. *Journal of Economic Theory* 39.

Banerji, A.V. and D. Cooper. 1991. 'Do Quantity-setting Oligopolists Play the Cournot equilibrium?'. Princeton University, Princeton. Mimeo.

Basu, K. 1990a. 'Duopoly Equilibria when Firms can Change their Decision Once'. *Economics Letters* 32.

——. 1990b. 'On the Non-existence of a Rationality Definition for Extensive Games'. *International Journal of Game Theory* 19.

Basu, K. and N. Singh, 1990. 'Strategic Entry-deterrence in Stackelberg Perfect Equilibria'. *International Economic Review* 31.

Benoit, J.P. and V. Krishna. 1985. 'Finitely Repeated Games'. *Econometrica* 53.

Binmore, K. 1987. Modeling Rational Players'. *Economics and Philosophy* 3.

Dixit, A. 1980. 'The Role of Investment in Entry-deterrence'. *Economic Journa1* 90.

Fraysse, J. and M. Moreaux. 1985. 'Collusive Equilibria in Oligopolies with Finite Lives'. *European Economic Review* 27.

Friedman, J. 1971. 'A Non-cooperative Equilibrium for Supergames'. *Review of Economic Studies* 38.

——. 1985. 'Cooperative Equilibria in Finite Horizon Non-cooperative Supergames'. *Journal of Economic Theory* 35.

Green, E. and R. Porter. 1984. 'Noncooperative Collusion under Imperfect Price Information'. *Econometrica* 52.

Harrington, J.E. 1987. 'Collusion in Multiproduct Oligopoly Games under a Finite Horizon'. *International Economic Review* 28.

Kreps, D.M., P. Milgrom, J. Roberts, and R. Wilson. 1982. 'Rational Cooperation in the Finitely-Repeated Prisoners Dilemma'. *Journal of Economic Theory* 27.

Pal, D. 1991. 'Cournot Duopoly with Two Production Periods and Cost Differentials'. *Journal of Economic Theory* 55.

Radner, R. 1980. 'Collusive Behaviour in Noncooperative Epsilon-equilibria of Oligopolies with Long but Finite Lives'. *Journal of Economic Theory* 22.

Rofemberg, J.J. and G. Saloner. 1986. 'A Supergame-theoretic Model of Price Wars during Booms'. *American Economic Review* 76.

Saloner, G. 1987. 'Cournot Duopoly with Two Production Periods'. *Journal of Economic Theory* 42.

Selten. R. 1978. 'The Chain-store Paradox'. *Theory and Decision* 9.

Slade. M.E. 1987. 'Interfirm Rivalry in a Repeated Game: An Empirical Test of Tacit Collusion'. *Journal of Industrial Economics* 35.

12 Stackelberg Equilibrium in Oligopoly

An Explanation Based on Managerial Incentives

12.1. INTRODUCTION

The aim of this chapter is to suggest a new explanation for Stackelberg leadership in a duopoly. There is a growing literature that theorizes about why there may be a Stackelberg outcome in some duopolies and which firm will be the leader (see, for example, Gal-Or 1985; and Boyer and Moreaux 1987). The present chapter shows that if owners of firms are free to delegate output decisions to managers appointed by them, then in the subgame perfect equilibrium a duopoly may equilibrate at a Stackelberg outcome. I proceed by adapting the models of Vickers (1984), Fershtman and Judd (1987), and Sklivas (1987) so as to allow owners of firms a larger and more realistic menu of choices.

In Fershtman and Judd, and Sklivas, attention is restricted to a class of linear contracts that owners may offer to the managers. After the owners have done so, each manager decides how much to produce. In these papers, however, the employer's decision to hire or not hire a manager is not explicitly modelled. In effect, the existing models look at the case in which the managers are already installed in the firms.

If, however, the decision to hire a manager is endogenous, then these models yield a very interesting possibility. The purpose of this chapter is to demonstrate that for a class of parameters, the subgame perfect-equilibrium strategies lead to production levels that are exactly equal to what would happen under

From *Economic Letters* 49, 1995: 459–64.

Stackelberg leadership. We then identify the conditions under which a particular firm becomes a Stackelberg leader.

12.2. The Model

In this section a related version of the Vickers (1984), Sklivas (1987), and Fershtman and Judd (1987) models is outlined. A duopoly faces the following inverse demand function:

$$p = a - b(x_1 + x_2),$$

where $a, b > 0$, p is price, and x_i is firm i's output. Firm i's cost of producing x_i units is given by $c_i x_i$. As usual, firm i's profit function and sales function are given by, respectively,

$$\pi_i = \pi_i(x_1, x_2) = (a - b(x_1 + x_2) - c_i)x_i, \tag{12.1}$$

and

$$S_i = S_i(x_1, x_2) = (a - b(x_1 + x_2))x_i. \tag{12.2}$$

In order to make the hiring decision explicit, let us suppose that in period 1 the *owners* of firms decide whether to hire *managers* or not. In particular, each owner i selects $m_i \in \{0, 1\}$, where $m_i = 0$ means owner i does not hire a manager and $m_i = 1$ means i does hire a manager. Once a manager is chosen and the manager's objective function specified, which happens in period 2, the manager decides how much to produce (in period 3). In the absence of a manager the decision (in period 3) is taken by the owner.

In period 2 each owner (who has a manager) picks an objective function for the manager. Manager i's objective function can only belong to the following class.

$$R_i = R_i(\alpha_i, x_1, x_2) \equiv \alpha_i \pi_i + (1 - \alpha_i)S_i. \tag{12.3}$$

In other words, owners 1 and 2 select α_1 and α_2, respectively, in period 2. Actually manager i is told that her salary is

$$A_i + B_i R_i,$$

where A_i and B_i are constants. Clearly, maximizing $A_i + B_i R_i$ and maximizing (12.3) are equivalent if the control variable is x_i. Hence, when speaking of managerial behaviour we shall speak as if the manager's objective function were $R_i(\alpha_i, x_1, x_2)$.

A_i, and B_i are chosen by owner i to simply ensure that the participation constraint is satisfied; that is, $A_i + B_i R_i$ in equilibrium happens to be equal to the manager's reservation income. Let us suppose that manager i's reservation income is Y_i. In addition, assume that the owner finds that he can get away from the firm for several hours once he has a manager. Let X_i be the amount that owner i can earn elsewhere in the time that gets released in this manner. Hence owner's i's cost of hiring a manager is

(12.4) $\qquad Z_i \equiv Y_i - X_i$

Owner *i's* *net* profit, if he hires a manager, will be $\pi_i - Z_i$.

It seems reasonable to assume $Z_i > 0$. This is because an owner cannot put all his time in another trade even if he has a manager. He will still have to do *some* supervision. What is interesting and at first sight counter-intuitive is that Stackelberg leadership can arise even when $Z_1 = Z_2$.

The model of Fershtman–Judd and Sklivas is a special case of the above game; namely, one in which in period 1, $(m_1, m_2) = (1, 1)$ is given. The subgame that occurs after this coincides exactly with their model and the subgame-perfect equilibrium of this *sub*game will be referred to as an 'incentive equilibrium'. The values taken by α_i, x_i, and π_i in such a subgame-perfect equilibrium are given by

(12.5) $\qquad \bar{\alpha}_i = (8c_i - a - 2c_j)/5c_i, \qquad j \neq i,$

(12.6) $\qquad \bar{x}_i = (2a - 6c_i + 4c_j)/5b, \qquad j \neq i,$

(12.7) $\qquad \bar{\pi}_i = 2(a - 3c_i + 2c_j)^2/25b.$

It follows from (12.3) that

(12.8) $\qquad x_i = \underset{x_i}{\operatorname{argmax}}\, R_i(\alpha_i, x_1, x_2) = (a - bx_2 - \alpha_1 c_1)/2b.$

I shall refer to (12.8) as the *managerial reaction function* of firm *i*. The owner's reaction function, referred to here as simply the *reaction function,* is clearly the special case of (12.8) with $\alpha_i = 1$. By solving the two equations in (12.8), with $i = 1, 2$. we get $x_1(\alpha_1, \alpha_2)$ and $x_2(\alpha_1, \alpha_2)$. By inserting these in (12.1) we get π_i as a function of (α_1, α_2) By treating this as the pay-off function of a normal-form game (where α_1 and α_2 are the strategic variables), and solving for the Nash equilibrium, we get $\bar{\alpha}_1$ and $\bar{\alpha}_2$. Then $\bar{\pi}_i$ in (12.7) is derived by replacing x_1 and x_2 with $x_1(\bar{\alpha}_1, \bar{\alpha}_2)$ and $x_2(\bar{\alpha}_1, \bar{\alpha}_2)$.

12.3. The Stackelberg Solution

The subgame-perfect equilibrium of the full game described in Section 12.2—namely, a game in which the hiring decision is an explicit one—however, need not coincide with the incentive equilibrium described in (12.5)–(12.7). For a certain class of parameters it would, in fact, coincide with the standard textbook Stackelberg equilibrium. In most conventional treatments, a Stackelberg solution is either *assumed* or directly deduced from an *exogenously* imposed temporal structure of moves. At times it is *motivated* by pre-entry configurations (as in, for example, Basu and Singh 1990) but it is left at the level of motivation. In the present model, the occurrence of a Stackelberg solution is more endogenous, and who will be leader depends on the cost parameters of the model.

To see this, let us first check what happens after each possible one-period history. We have already seen what happens after $(m_1, m_2) = (1, 1)$. Next, consider the case $(m_1, m_2) = (0, 0)$; that is, no one hires a manager. This gives us the standard Cournot case and as is easily worked out or, even more easily, lifted from some microeconomics textbook, firm *i*'s profit in the Cournot equilibrium is given by

(12.9) $\pi_{iN} = (a - 2c_i + c_j)^2/9b.$

Finally, and without loss of generality, consider the history $(m_1, m_2) = (0, 1)$. This case is solved in Fershtman (1985). Since $m_1 = 0$, firm 1's reaction in period 3 is the owner's reaction function, given by setting $i = 1$ and $\alpha_1 = 1$ in (12.8). This is described by *AB* in Figure 12.1.

Let *CD* be owner 2's reaction function. Since, by choosing $\propto_2$, owner 2 can make any line parallel to *CD* firm 2's managerial reaction function, it is obvious that in period 2 owner 2 would choose α_2 such that the managerial reaction function is *C″D″*, which is a line that goes through point *S*, which is a standard Stackelberg outcome with firm 2 as leader. The line π_2'' is firm 2's iso-profit curve.

The profit earned by firm 1 at point *S* in Figure 12.1 is denoted by π_{1F}. The subscript is explained by the fact that firm 1 is a 'follower' at point *S*. The computation of π_{1F} is standard:

(12.10) $\pi_{1F} = (a - 3c_1 + 2c_2)^2/16b,$

π_{1F} is symmetric.

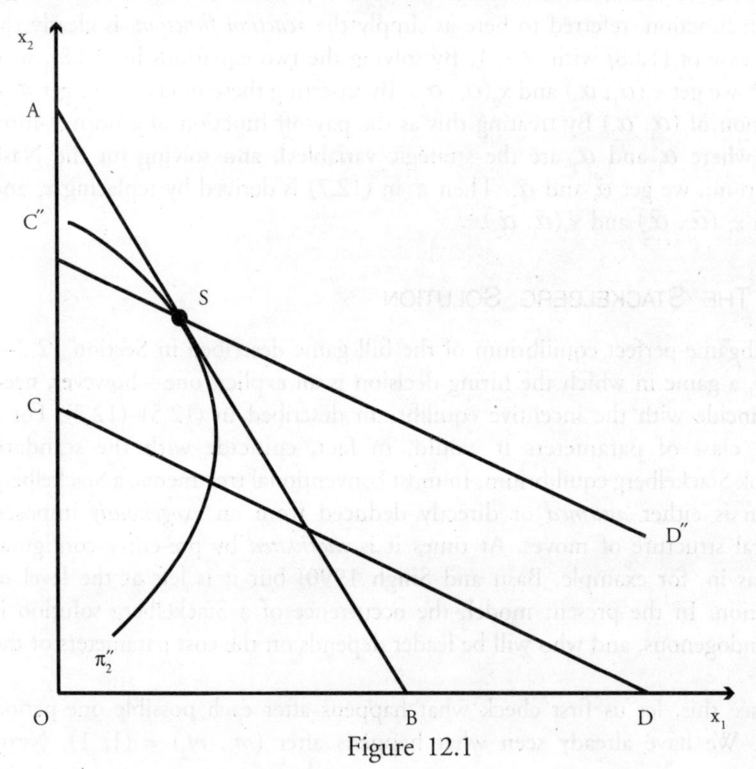

Figure 12.1

If π_{1L} is 1's profit when 1 is Stackelberg leader, it is easy to check that

(12.11) $$\pi_{1L} = (a - 2c_1 + c_2)^2/8b.$$

Recalling that hiring a manager costs firm 1, Z_1, owner 1's pay-offs that occur in the subgame-perfect equilibria of the subgames that occur after the four possible period-1 histories, (0, 0), (1, 1), (0, 1), (1, 0), are summarized in the following partial pay-off matrix:

		Owner 2	
		0	1
Owner 1	0	π_{1N}, π_{2N}	$\pi_{1F}, \pi_{2L} - Z_2$
	1	$\pi_{1L} - Z_1, \pi_{2F}$	$\bar{\pi}_1 - Z_1, \bar{\pi}_2 - Z_2$

The incentive equilibrium is a subgame-perfect equilibrium of the full game only if $\pi_1 - Z_1 \geq \pi_{1F}$ and $\bar{\pi}_2 - Z_2 \geq \pi_{2F}$. It is easy to work out these conditions in terms of the exogenous parameters of the model. If, on the other hand,

(12.12) $$\pi_{1F} > \bar{\pi}_1 - Z_1 \text{ or, equivalently, } Z_1 > \frac{7(a - 3c_1 + 2c_2)^2}{400b},$$

and

(12.13) $$\bar{\pi}_{2L} - Z_2 > \pi_{2N} \text{ or, equivalently, } \frac{(a - 2c_2 + c_1)^2}{72b} > Z_2,$$

then a subgame perfect equilibrium of this model coincides with the standard Stackelberg outcome with 2 as leader.

From (12.12) and (12.13) it follows that the equilibrium is likely to coincide with the Stackelberg outcome with firm 2 as leader if (i) firm I's marginal cost of production is high, relative to firm 2's marginal cost and (ii) Z_1 is high relative to Z_2.

Since Z_1 and Z_2 are the costs of hiring a manager. it may at first sight appear that the Stackelberg result is being driven by (ii). However, surprisingly, even if $Z_1 = Z_2 = Z$, (12.12) and (12.13) can be satisfied. All we need is that

(12.14) $$\frac{7(a - 3c_1 + 2c_2)^2}{400b} < Z < \frac{(a - 2c_2 + c_1)^2}{72b}$$

And this can be true if c_1 is sufficiently high compared with c_2. Note, however, that if $Z = 0$, then (12.14) cannot be true and, therefore, that Stackelberg cannot occur.

Notes

1 See: also, D'Aspremont and Gerard-Varet (1980); Fershtman (1995); and Katz (1986)

References

Basu, K. and N. Singh. 1990. 'Entry-deterrence in Stackelberg Perfect Equilibria'. *International Economic Review* 31.

Boyer, M. and M. Moreaux. 1987. 'Being a Leader or a Follower: Reflections on the Distribution of Roles in Duopoly'. *International Journal of Industrial Organization* 5.

D'Aspremont, C. and L.-A. Gerard-Varet. 1980. 'Stackelberg-solvable Games and Pre-play Communication'. *Journal of Economic Theory* 23.

Fershtman, C, 1985. 'Internal Organization and Managerial Incentives as Strategic Variables in a Competitive Environment'. *International Journal of Industrial Organization* 3.

Fershtman, C. and K.L. Judd. 1987. 'Equilibrium Incentives in Oligopoly'. *American Economic Review* 7.

Gal-Or, E. 1985. 'First Mover and Second Mover Advantages'. *International Economic Review* 26.

Katz, M.L. 1986. 'Game-playing Agents and Unobservable Contracts'. Princeton University. Mimeo.

Sklivas, S.D. 1987 'The Strategic Choice of Managerial Incentives'. *RAND Journal of Economics* 18.

Vickers, J. 1984. 'Delegation and the Theory of the Firm'. *Economic Journal* (Supplement) 95.

13 The Strategic Role of International Credit as an Instrument of Trade

with Ashwini Deshpande

13.1. INTRODUCTION

The use of international credit by the donor country to boost its exports and exports earnings has been extensively discussed in the literature on political economy and in the recently emerging literature on loan pushing.[1] Indeed, this was one of the tenets of the critique of the industrialized-country behaviour vis-à-vis the Third World in several radical writings and is also often used by conservative policy makers in the North to justify the giving of aid. The objective of this chapter is to examine this thesis rigorously and by using standard techniques of modern microeconomic theory. It turns out that the relation between credit and exports is more complicated than appears at first blush.

Since most Third World nations have non-convertible currencies, when they receive a loan from the rich nation in the rich nation's currency, they will use the loan for the direct buying of goods from the donor country. If the rich donor country's currency is easily convertible with another rich country's currency, the direct demand would be generated for both these countries' goods. So there may be some externality in demand. Donor nations have often used tied-in credit to minimize such externality and to ensure that each country's credit would be used for their own products.[2] So indeed there is a

From The Japanese Economic Review, 4, December, 1995: 333–50.

We are grateful to Jonas Bjornerstedt, Avinash Dixit, Ashok Guha, Henrik Horn, Dilip Mookherjee, Bob Rowthorn, Partha Sen, Nick Stern, and Eric Thorbecke for comments and suggestions. We also benefited from discussions at seminar presentations of this paper at the Indian Statistical Institute, New Delhi, Stockholm University, the London School of Economics, the University of Cambridge and Cornell University.

prima facie case for believing that there is a close link between international credit and donor exports. However, the nature of this relation needs careful exploration.

The link between international credit and international trade is not a recent phenomenon; it has been argued that such a relationship can be traced back to the colonial period. For instance, Rothermund (1981), while analysing the British trade policy in India during the Great Depression, concludes that the British evolved a package of measures (like imperial preferences, a high exchange rate of the rupee that acted as an import bonus, and so on) to ensure their dominance in Indian markets. Similarly, in the context of US aid, in particular that of aid disbursed by the Agency for International Development (AID), Hyson and Strout (1968) observe how by 1966, as a consequence of aid-tying policy, '\$9 out of every \$10 of foreign commodity expenditures financed by AID went to US suppliers'.

Much writing on aid and aid policy, by radical writers and others,[3] argued that while aid was ostensibly given out of humanitarian concerns, the actual motives were considerations of political, military, and economic advantage for the donors. Ohlin (1966) reports that in 1963, 84 per cent of development assistance was bilateral (rather than multilateral) and argues that tied aid was inspired by hopes of returns—whether in the form of political advantage or economic advantage to donors. Tendler (1975) points out that the availability of aid exclusively for foreign exchange costs of a project also resulted in a larger problem, namely, the priorities of the recipient country invisibly rearranged themselves around foreign-exchange intensive projects. In the 1970s, aid gradually declined in importance and international credit took the form of private loans from the large money centre banks in the advanced industrialized countries.

The purpose of this chapter is to develop a model which can be used for analysing the link between this form of international credit and trade. It turns out that the relation between credit and trade is much more varied than has been believed by earlier writers. 'Pathologies' can occur easily. For instance, making more credit available to a developing country can cause a shrinkage in the quantity of donor exports to the recipient country. It can also cause a fall in the profits of the exporting firms.

The models developed in this chapter deliberately stay away from theoretical problems concerned with the role of money. It is but natural that in studying the links between international credit and trade some of the same problems of money-in-general-equilibrium will arise. Those are deep problems and for a chapter like the present one the best strategy is to look away from them. So we begin by assuming that a poor country needs hard currency or 'foreign' exchange to purchase goods from an industrialized country and that its own hard currency reserves are inadequate.

Though it is a well-known fact that most less-developed countries (LDCs) cannot pay for their imports with their own currency but must pay with some hard currency, it may be questioned why this is so. While we do not wish to enter into a detailed analysis (and we are content to treat this 'fact' as an axiom throughout this chapter), it may be pointed out that a clue to the answer may lie in asymmetric information.

It is arguable that if India buys goods from Japan and pays for these in Indian rupees, India would have a better idea of what the rupees will be able to buy than the Japanese. However, if India pays in US dollars, it is not clear that India would be better placed than Japan to know what the dollars can buy. We should, however, clarify here that our formal model is not dependent on the convertibility assumption. An alternative assumption which would suffice is that the foreign exchange available with the South is explicitly-tied credit, that is, money which the South *has* to spend on the North's products.

In Section 13.2 the method of our analysis is introduced by considering the simple case of a monopolist exporter in an industrialized country. The effect on his profit and sales of making credit available to a poor country is analysed and nothing surprising happens here. It merely serves as a benchmark for the rest of the chapter. It may be noted that all the models developed in this chapter are partial-equilibrium, microeconomic models. In other words, this chapter ought not to be treated as an exercise in international trade which is usually cast in general-equilibrium terms.

Section 13.3 shows that 'pathologies'—indeed it is not clear that these can be called pathologies—can easily arise in the relation between exports and profits, on the one hand, and international credit, on the other, once we allow for oligopoly among exporters.[4] In Section 13.4, the donor country bank is introduced as a strategic profit-making agent which first decides how much to lend and at what terms; then the exporters move into action. The subgame perfect equilibrium of this game is characterized. We also find that there exist states in which the recipient country is worse off as a result of international credit being made available to it to finance its imports of manufactured goods. We call this the 'Adverse Effect Theorem' and given its counter-intuitive nature it is a theorem of some significance. It is a standard belief that even if international credit is used by the donor for his own benefit, from the point of view of the Third-World recipient nation it is always better to have credit available than not. Pincus (1963), for instance, writes, 'In practice, the United States and most other donors do tie their aid, and this custom, whatever its disadvantages, is clearly preferable to no aid at all'. What our theorem shows is that this claim is not valid.

The proof of this theorem occurs in Section 13.5. A few concluding remarks are offered in Section 13.6.

13.2. Credit and Exports: A First View

Throughout this chapter we restrict attention to a two-country world. The North is an industrialized nation, exporting a manufactured good produced by large firms; its currency is called the hard currency. The South is an underdeveloped country,[5] where many consumers buy or (wish to buy) the manufactured good from the North. Its currency is called the soft currency. A Northern producer will refuse to sell his goods against soft currency; he insists on being paid in hard currency. This could be because of the exchange rate being pegged at a non-market clearing level or expectations of uncertainty in the South or informational asymmetries of the kind discussed in Section 13.l.

The South suffers from a shortage of hard currency. Suppose the South possesses R units of hard currency. What we mean by a shortage is that if the South had more hard currency, it would have demanded more manufactured goods. Since this weakness of demand is caused by a shortage of hard currency and not by a shortage of money in general (i.e., income), what is implicitly being assumed is that in the future the South expects to have adequate foreign exchange or equivalently to have a freely convertible currency. So if it can get a (hard currency) loan now it would be able to pay it back later.

To formalize this, suppose that the South's demand function for the manufactured good from the North, assuming that it has no foreign-exchange constraint, is given by:

$$x = x(p), \tag{13.1}$$

where $x(p)$ is a usual, downward-sloping demand curve. The inverse of this is given by:

$$p = p(x). \tag{13.2}$$

However, this demand function is not necessarily the effective one because all foreign goods have to be bought in hard currency and the South has only R units of hard currency, Let the exchange rate be e.[6] That is, each unit of hard currency can be converted to e units of soft currency. Hence, since price, p, is given in soft currency, the actual demand function is given by:

$$x = \min\{x(p), eR/p\} \tag{13.3}$$

This demand function is illustrated in Figure 13.1. The thick line is the actual demand curve.

A word clarifying possible assumptions underlying the demand function in (13.3) is useful. We could either assume that the borrowing country has one consumer who is nevertheless a price-taker (somewhat in the manner of the price-taking agents we depict in the two-person Edgeworth Box) or that the borrowing country has many identical price-taking agents and the limited foreign exchange is distributed equally among the consumers. Either of these assumptions would justify using (13.3). An interesting future exercise could be to develop models of the kind presented here but with different rules for rationing out the limited foreign exchange available with a developing country. This could help determine what the 'best' rationing rule is from the point of view of the developing nation.

Let us, as a benchmark, see what would happen if the Northern manufacturing industry was a monopoly. Let the monopolist's cost of production be linear with per-unit cost fixed at ec in soft currency units. To see where the equilibrium

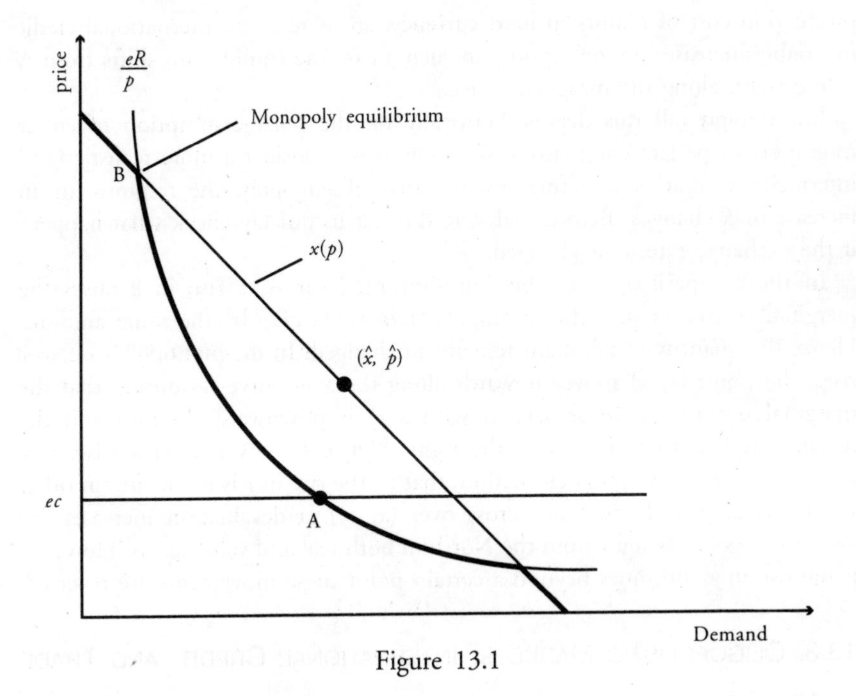

Figure 13.1

will be, let us, initially, work out the monopoly equilibrium assuming that the foreign-exchange constraint does not exist. In that case the amount sold by the monopolist is

$$\hat{x} = \text{argmax } [p(x) - ec]x,$$

and the equilibrium price is $\hat{p} = p(\hat{x})$.

Now to characterize the monopolist's equilibrium, when he confronts the actual demand curve, (13.3), is simple. Check if at $(\hat{x}, \hat{p})$ the foreign-exchange constraint is binding or not. Consider first the case when it is not binding, that is, $\hat{x}\hat{p} < eR$. In that case, $(\hat{x}, \hat{p})$ is a point on the thickened part of the $x(p)$ curve in Figure 13.1. Since that is feasible, that is the monopoly-equilibrium point. Next, suppose $\hat{x}\hat{p} > eR$. This is when the foreign-exchange constraint is binding. This case is illustrated in Figure 13.1. In that case [given that (13.3) is the demand curve], the marginal revenue curve will have a discontinuity below point B [which marks the left-hand intersection of $x(p)$ and eR/p]. Thereafter, marginal revenue coincides with the horizontal axis. Hence the monopoly equilibrium occurs at B.

The impact on exports of giving international credit to the South is now easy to study in the case where equilibrium is at B, to start with. If a loan of L units of hard currency is given, the curve eR/p will shift to the right. The new curve will be $e(R + L)/p$. Hence, the monopoly-equilibrium point will move down the $x(p)$ curve. In other words, the Northern exports will rise, both in real and value terms and the price of the good will fall. This is as would be expected. However, this proposition hinges critically on the market structure being that of a monopoly. Also, in the polar case of perfect competition, where all firms have a production cost of c units in hard currency, an increase in international credit invariably increases donor exports. In such a case the equilibrium shifts from A to the right, along the marginal cost curve.

Interestingly, all this depends critically on the polar assumption of either monopoly or perfect competition. As soon as we consider a more realistic (and intermediate) market structure, to wit, that of oligopoly, the response to an increase in R changes. Before analysing this, let us quickly check what happens if the exchange rate, e, is changed.

In the competitive case when equilibrium is at A, a rise in e raises the marginal cost curve and the rectangular hyperbola eR/p by the same amount. Hence the quantity of x bought remains unchanged. In the monopoly case, as e rises, the point $(\hat{x}, \hat{p})$ moves upwards along the $x(p)$ curve (assuming that the marginal revenue curve associated with $x(p)$ is downward sloping) and the rectangular hyperbola moves to the right. Hence B moves downwards along $x(p)$. Therefore, as long as devaluation, that is, the rise in e is moderate (in other words, as long as B does not cross over $(\hat{x}, \hat{p})$), a devaluation increases the amount of goods bought from the North in both real and value terms. However, if the rise in e continues beyond a certain point these movements are reversed.

13.3. Oligopolistic Markets, International Credit, and Trade

In this section we consider the case where there are two exporting firms in the North. The generalization to the case of n firms is trivial. These firms are Cournot oligopolists. They choose quantities and the price is determined by the market. We do not bring the domestic market of the North into the model not because these products are not sold in the North but because marginal costs are constant and price discrimination between countries is assumed to be possible. This separates the markets of the North and the South (see, for example, Dixit 1984) and allows us to focus exclusively on each of these. Before characterizing the Cournot equilibrium for a model where the demand curve is given by (13.3) above, we make some simplifying assumptions; and for this it is convenient to introduce some new terms.

If the oligopolists face a demand curve given by eR/p, we shall refer to the Cournot equilibrium as the R-Cournot equilibrium; and, similarly, the reaction functions, iso-profit curves, etc., as R-reaction functions, R-iso-profit curves, etc. If on the other hand, the demand curve were $x(p)$, then the equivalent concepts will be referred to as the x-Cournot equilibrium, x-reaction functions and x-iso-profit curves. When we talk of the real problem, that is, with the demand curve given by (13.3), then we simply drop the letters R and x from the above terms.

It will be assumed throughout that there is a unique x-Cournot equilibrium, that each firm's x-profit function is strictly concave and the x-reaction functions are downward sloping.

In setting out to characterize the Cournot equilibrium of the model it is useful to first describe the R-Cournot equilibrium. Note that if the demand function faced by the oligopolists is given by eR/p, then each firm's profit function is:

$$(13.4) \qquad \pi_i(x_1, x_2) = [eR/(x_1 + x_2) - ec]x_i,$$

where x_j is the output produced by firm $j \in \{1, 2\}$.

The first-order condition for maximizing π_i, is given by

$$\partial\pi_i/\partial x_i = eR/(x_1 + x_2) - x_i eR/(x_1 + x_2)^2 - ec = 0.$$

To check the second-order condition note that

$$\partial^2\pi_i/\partial x_i^2 = [2eR/(x_1 + x_2)^2][x_i/x_1 + x_2) - 1] < 0.$$

The first-order condition above implies the following:

$$(13.5) \qquad x_i = \sqrt{Rx_j/c} - x_j, \qquad j \neq i$$

If (x_1^R, x_2^R) satisfies (13.5) for $i = 1, 2$, then (x_1^R, x_2^R) is an R-Cournot equilibrium.

It is easy to see that an R-Cournot equilibrium is unique. Note that (13.5) implies that

$$x_i + x_j = \sqrt{Rx_j/c} \qquad j = 1, 2$$

or

$$x_1 + x_2 = \sqrt{Rx_j/c} \qquad j = 1, 2.$$

Hence $x_1^R = x_2^R$.

Let us denote x_1^R by x^R. Equation (13.5) implies that

$$x^R = R/4c.$$

It follows that the equilibrium price, p^R, is:

$$p^R = eR / 2x^R = 2ec.$$

Observe that equilibrium price does not depend on R. Hence as the foreign-exchange reserves with the South rise from R^0 to R^1, the R-equilibrium point shifts from E^0 to E^1 in Figure 13.2.

Next we need to establish a lemma on the taxonomy of equilibria with intersecting demand curves. Let us work this out in general terms before relating it to our exercise in hand.

Let $x = y(p)$ and $x = z(p)$ be two downward-sloping demand curves. Consider separately the cases where our duopolists confront each of these as the aggregate demand curve. Following in an obvious way, the terminology introduced earlier, let us assume that the Cournot equilibrium is unique in each of the y-case and the z-case, that it is stable, that each firm's y-profit function and z-profit function are strictly concave and that the y-reaction function and the z-reaction function of each firm is downward sloping. Suppose now that the two firms confront the demand curve $\phi(p) = \min\{y(p), z(p)\}$. Given the alternative descriptions of the y-Cournot and z-Cournot equilibria in Figure 13.3, we wish to characterize the Cournot equilibrium or to be more explicit, the ϕ-Cournot equilibrium.[7]

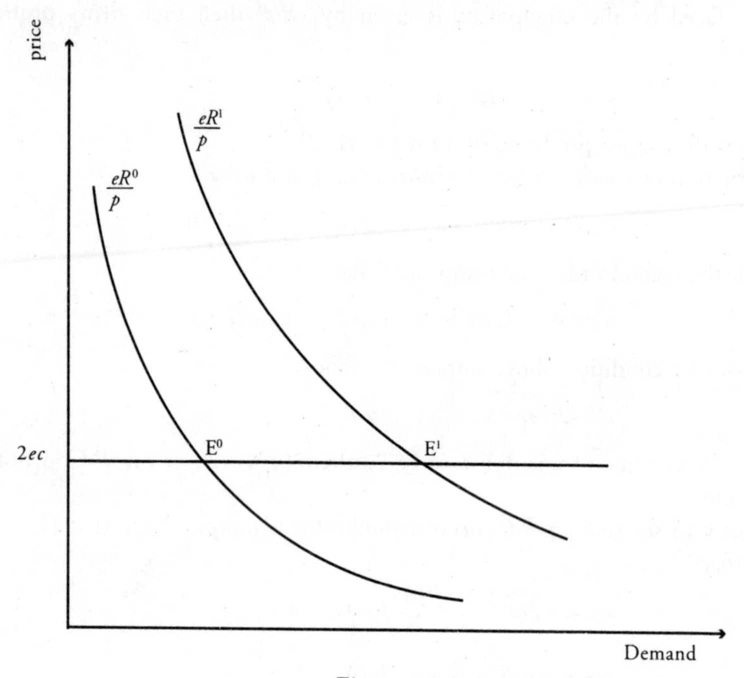

Figure 13.2

Let us first take up the case where the y- and z-Cournot equilibria are located as shown in Figure 13.3(a). It can be shown that in this case there is a unique ϕ-Cournot equilibrium and that this coincides with the z-Cournot point. To prove this, first check that the z-Cournot point is indeed a ϕ-Cournot point. To see this, suppose that the two firms are playing the ϕ-duopoly game but producing the z-Nash equilibrium output, (x_1^z, x_2^z). Thus price is given by p^Z. These are illustrated in Figure 13.3(a). Without loss of generality, consider deviations by firm 1. If this produces more, they move down along the $z(p)$ curve. Since such a move was not worthwhile for firm 1 in the z-case [recall that (x_1^z, x_2^z) is a Nash equilibrium], it cannot be worthwhile in the ϕ-case. What if firm 1 cuts back production? For a small cut back they move up along $z(p)$ and this cannot increase firm 1's profit. For a large cutback they begin to move up from A, along $y(p)$. This cannot increase firm 1's profit because moving up along $z(p)$ does not increase profit and moving up along $y(p)$ gives less profit than points vertically above $y(p)$ and on $z(p)$.

Next we have to show that a point like B cannot be a ϕ-equilibrium. This is easily done by using an argument similar to the above one and premised on the fact that B is not a y-Cournot point.

It is a little more complicated to show that A is not an equilibrium. Assume the two firms are producing outputs so that the aggregate price–output configuration is at A. The aggregate output here is less than the aggregate output at the z-Cournot point. Since the z-reaction functions are downward sloping and z-profit functions are strictly concave, it follows that at least one of the two firms

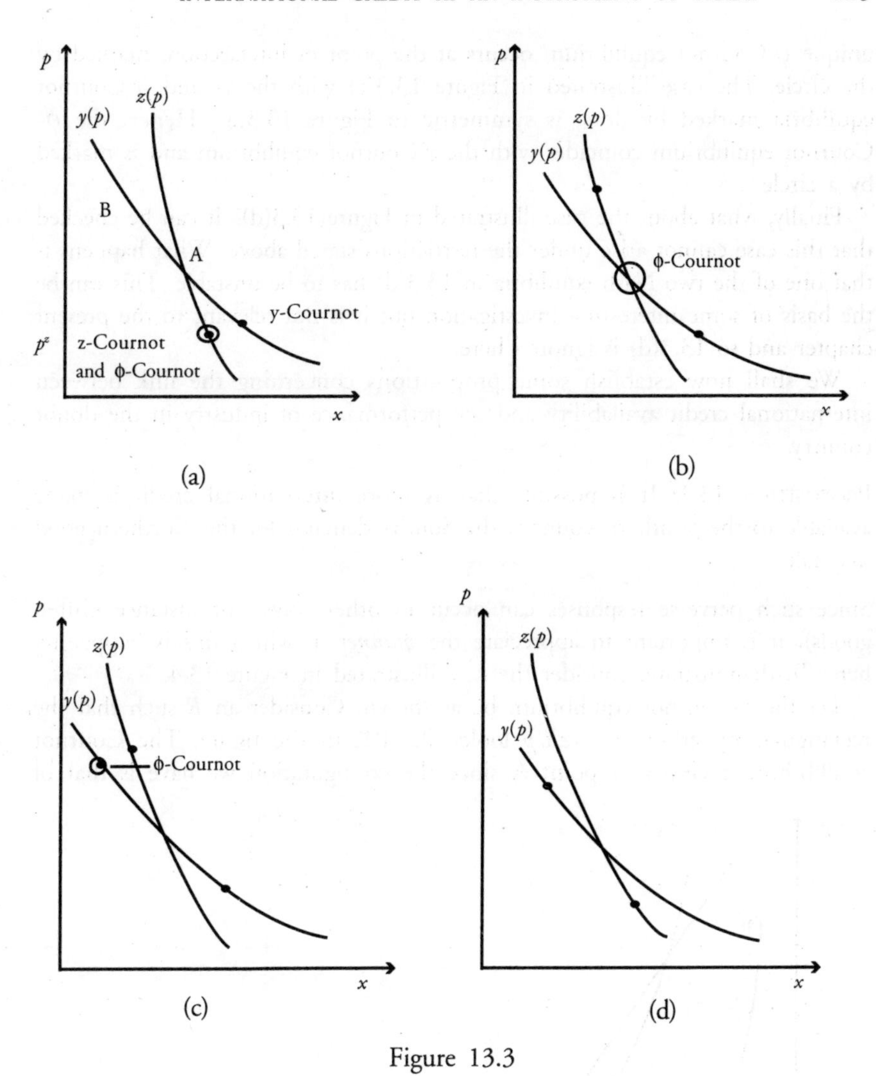

Figure 13.3

can increase its profits by increasing production. Hence A is not a ϕ-Cournot equilibrium. This establishes that the ϕ-Cournot equilibrium is unique and coincides with the *z*-Cournot equilibrium.

Consider now the case illustrated in Figure 13.3(b) where the *z*- and *y*-Cournot equilibria are marked by dots. Observe that neither of these dots is on the demand curve. Hence neither can qualify as the ϕ-Cournot equilibrium. Now consider any point on the $y(p)$ curve above the point of intersection marked by a circle. A duopoly will tend to move away from such a point since the *y*-Cournot equilibrium occurs elsewhere. However, the duopoly cannot proceed past the intersection point, since beyond this point the $y(p)$ curve ceases to be the demand curve. For a similar reason, the ϕ-Cournot equilibrium cannot occur on the $z(p)$ curve, below the intersection. It follows that the unique ϕ-Cournot equilibrium occurs at the point of intersection, marked by the circle. The case illustrated in Figure 13.3(c) with the *z*- and, *y*-Cournot equilibria marked by dots, is symmetric to Figure 13.3(a). Hence, the ϕ-Cournot equilibrium coincides with the *y*-Cournot equilibrium and is marked by a circle.

Finally, what about the case illustrated in Figure 13.3(d)? It can be checked that this case cannot arise under the restrictions stated above. What happens is that one of the two Nash equilibria in 13.3(d) has to be unstable. This can be the basis of some interesting investigation but it is not relevant to the present chapter and so 13.3(d) is ignored here.

We shall now establish some propositions concerning the link between international credit availability and the performance of industry in the donor country.

Proposition 13.1: It is possible that as more international credit is made available to the Southern country, the South's demand for the Northern good may fall.

Since such perverse responses can occur in other ways (for instance Giffen goods), it is important to appreciate the *manner* in which this is happening here. To demonstrate, consider the case illustrated in Figure 13.4.

Let the *x*-Cournot equilibrium be as shown. Consider an *R* such that the rectangular hyperbola $x = eR/p$ looks like RR in the figure. The Cournot equilibrium is clearly at point A since the configuration we have is that of

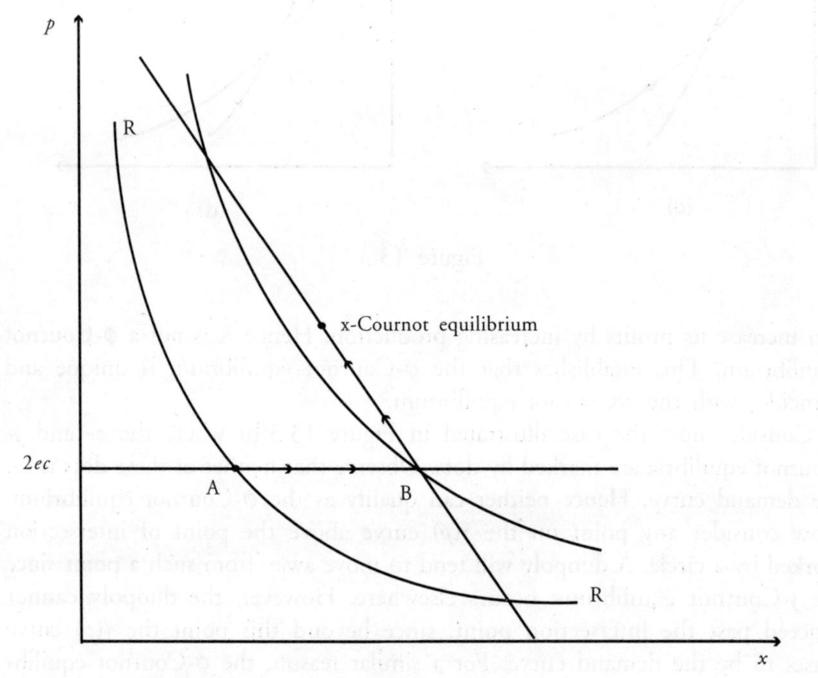

Figure 13.4

Figure 13.3(c). As R increases, the Cournot equilibrium moves horizontally as shown by the arrow. This happens up to point B. From here onwards, further increases in R create the configuration in Figure 13.3(b) and so, as R increases, the Cournot-equilibrium point begins to move up along the $x(p)$ curve as shown. This is the region where the claim in Proposition 13.1 occurs. In this region, if the industrialized country wishes to push exports on to the developing country it will be in its interest to limit the foreign-exchange reserves of the developing country!

If the R-constraint is binding, we have an equilibrium which is best described as one of *implicit excess demand*, because though there is no open excess demand, any increase in R translates into increased demand.

In the context of Proposition 13.1 above, it is of course arguable that an industrialized nation's primary objective is not the *quantity* of exports but profitability. Our next proposition pertains to this.

PROPOSITION 13.2: There exist situations where making more international credit available with the South lowers the profit of the Northern manufacturing industry.

The proof of Proposition 13.2 is obvious from Figure 13.5. As R increases, starting from R', equilibrium moves from A, through B, towards the x-Cournot equilibrium. This can be checked by comparison with Figure 13.3. Up to point B we have a situation like that in Figure 13.3(a). Then onwards we get the case illustrated in Figure 13.3(b). As the equilibrium traverses from A through B

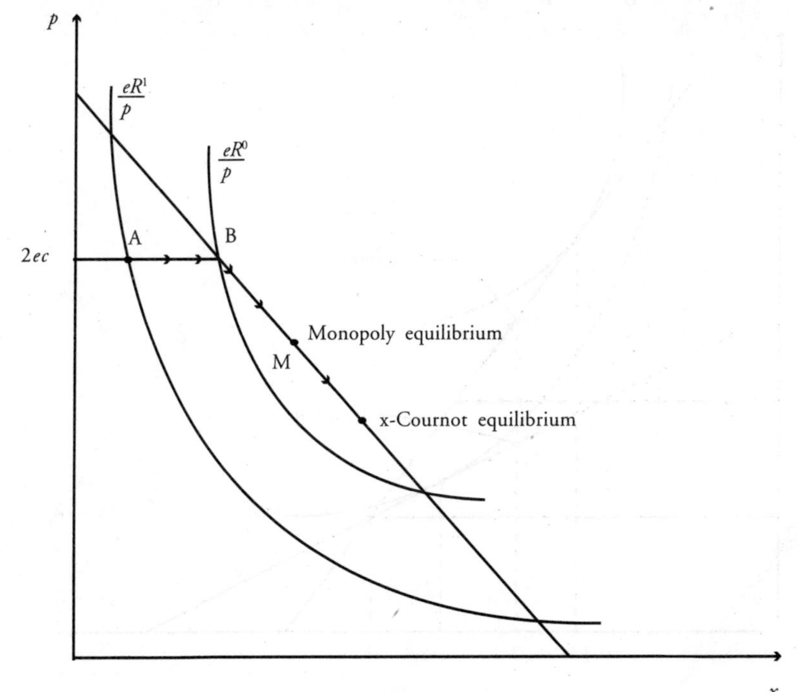

Figure 13.5

towards the x-Cournot equilibrium, clearly profit rises up to point M and then declines.

Note that if the industrial-sector firms have influence over their government's credit-giving institution, they will try to control the credit and use it as a device to help establish a collusive outcome. They will then ideally give credit so that the total foreign exchange with the South is just right to sustain a monopoly equilibrium. Hence, the government in the North could act as a collusion-facilitating device ensuring, through the suitable control of credit, that the Northern manufacturers earn monopoly profits from the South.

It may be useful to end this section with an example which illustrates the two propositions.

Assume, $e = c = 1$ and inverse demand (13.2) is given by:

$$x = 3 + B - Bp.$$

Check that at the x-Cournot equilibrium,

$$x_1 = x_2 = 1.$$

$$p = (1 + B)/B .$$

At the R-Cournot equilibrium, $p^R = 2$.

The reader may check that the critical value of B is 1. If $B < 1$, Proposition 13.1 is true. If $B > 1$, Proposition 13.2 is true. The cases of $B = 1/2$ and $B = 2$ are illustrated in Figure 13.6.

It is time now to go a step further and instead of varying the international

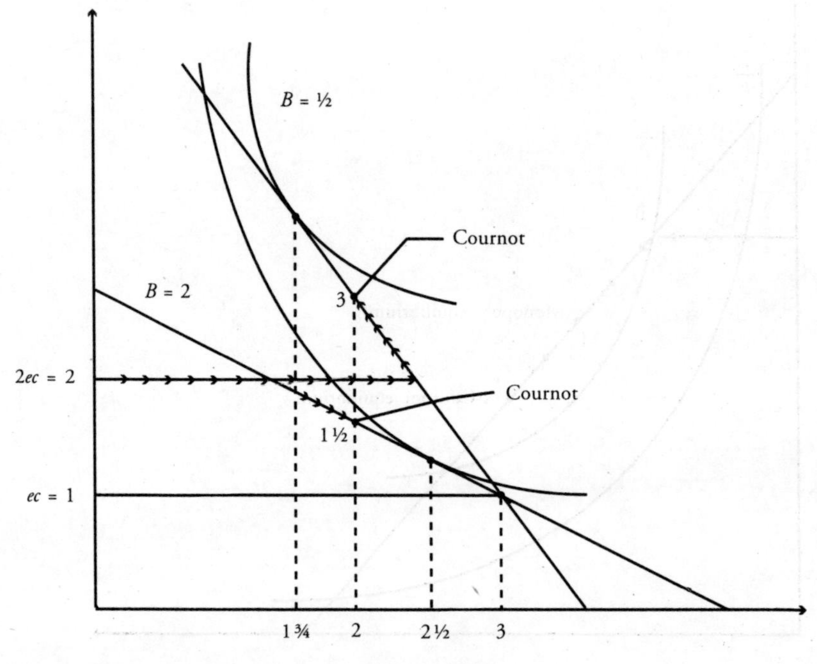

Figure 13.6

credit as if it were an exogenous variable we now model the donor country bank and endogenize the amount of credit given to the South.

13.4. The Equilibrium Amount of International Credit

In this section we consider a two-period model in which in period 1 the Northern bank decides how much credit to give or what interest rate to charge and in period 2 the Northern firms sell their products in the Southern market. We analyse the subgame-perfect equilibrium of this interaction.

Assume for the time being that every time goods are bought by the South from the North, the Southern government has first to borrow hard currency from the Northern bank at a *net* interest rate of i. If this is passed on to the Southern consumers, then when the price of the good is p, the effective price (or consumer price) faced by the Southern consumer is $(1 + i)p$. Hence, the demand function is given by:

$$x = x((1 + i)p) \tag{13.6}$$

That is, if international credit always had to be acquired by paying a premium, i, over and above the price, then we could use (13.1) but the argument would be $(1 + i)p$, as in (13.6), instead of p, as in (13.1). The interest, i, enters in the same way as an *ad valorem* tax.

The amount of money the South actually has to borrow is given by:

$$L = \max \{0, (px/e) - R\}.$$

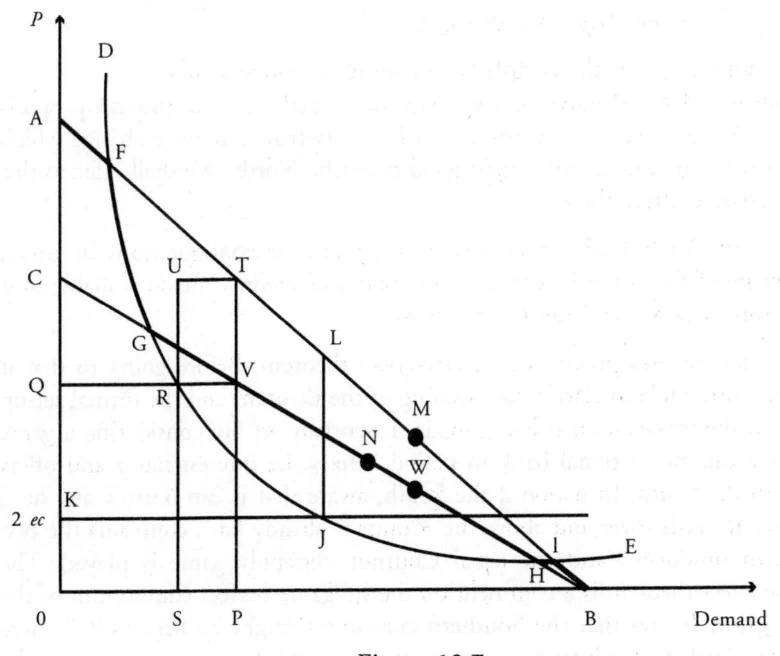

Figure 13.7

Note that if R was so large that the South did not have to borrow any money from the Northern bank, then the demand function for x would be given by (13.1) above. So (13.1) is the demand for manufactures when there is no foreign-exchange constraint. We shall in this section assume that the income effect associated with this good is zero. This is simply to enable us to do welfare analysis later without having to contend with several competing notions of consumer's surplus.

Comparing (13.6) and (13.1) it is clear that if AB, in Figure 13.7, is the x-demand function [i.e. (13.1)], then in the case where foreign exchange has to be borrowed at an interest of i to buy the good the new demand curve will be like CB, the vertical gap between the two demand curves being ip. We shall refer to this as the i-demand function. If i rises, CB will pivot at B and become flatter. The Cournot equilibrium on this demand curve will be called the i-Cournot equilibrium.

Recall that the demand function (13.6) (and therefore also CB) was derived under the assumption that $L > 0$. But if the amount spent on x is less than eR, then there is no need to borrow money, and $L = 0$. To take this into account superimpose the rectangular hyperbola represented by $x = eR/p$ on Figure 13.7. This is shown by the curve DE in Figure 13.7. To the right of the rectangular hyperbola DE, $L > 0$ and the demand curve is given by the relevant part of CB. Not so to the left of DE.

Hence, the overall demand function that emerges is given by the thickened line, AFGHIB. In terms of algebra, the demand function we are now considering is given by:

$$x = \text{mid}\,\{x(p),\ eR/p,\ x((1+i)p)\}, \quad (13.7)$$

where mid $\{a, b, c\}$ is the middle largest number among a, b, c.

With this demand curve that we have developed,[8] we can prove a proposition about the welfare of the South if it has to borrow foreign exchange which enables it to buy the manufactured good from the North. We shall refer to this as the adverse effect theorem.

PROPOSITION 13.3: There exists a class of parametric configurations in which the welfare of the South is reduced if international credit is made available to it by a profit-maximizing international bank.

The next section proves the adverse-effect theorem. Before going to that it may be worthwhile to clarify the meaning of the theorem and the formal set-up in which this observation is being made. Essentially we are considering a game in which the international bank in period 1 fixes the interest rate i and offers credit to the South. In period 2 the South, aware that it can borrow any hard currency it needs (over and above the R units it already has), confronts the two Northern producers, and the usual Cournot oligopoly game is played. The adverse-effect theorem is a comment on the subgame perfect equilibrium of the above game. It says that the Southern consumers might be better off if there were no Northern bank lending money to them and they were forced to make do with their existing foreign-exchange balance, R, than in the subgame perfect equilibrium of the above game involving the international bank.

In the next section we also try to demonstrate that the adverse-effect theorem is not just a logical possibility but may well arise in natural situations. This is demonstrated at the end of the next section with an example and suggestions of how this may be generalized to a *class* of examples.

13.5. Proof of the Adverse-effect Theorem

To prove the adverse-effect theorem, begin by inspecting Figure 13.7. The new demand curve has been formed through the juxtaposition of three separate demand curves: the x-demand curve, the i-demand curve, and the eR/p curve (the rectangular hyperbola). Using the terminology developed earlier, let the x-Cournot equilibrium be given by point M and the R-Cournot equilibrium by point J.

If the demand curve that the South faces is given by (13.3), that is, min $\{x(p), eR/p\}$, then (as proved in Section 13.3) the Cournot equilibrium will be given by J, and the consumers' surplus of the South will be KJLA. This is a situation where the South does not borrow in order to buy the manufactured good. However, as has been established earlier, if the South has to take recourse to international credit, the demand curve that it faces is AFGHIB. Thus, to prove our contention, we need to prove that the consumers' surplus in the latter case can be smaller than KJLA.

From Figure 13.7 it is clear that this will certainly be the case if the Cournot equilibrium on the new demand curve (i.e. AFGHIB) is to the left of point J, say at V. At V, the foreign exchange that the South has is sufficient to enable it to buy only QR units of the good at the market price (OQ). Thus to buy the remaining quantity, it borrows at an interest rate i with demand being given by CB. The consumers' surplus here is smaller for two reasons: the South is buying a smaller quantity at a higher price and the price now also includes the interest payment to the bank, with the consumers' surplus being reduced by the amount of profit that the bank makes. By making a loan of RVPS, the profit of the bank will be RVTU.[9] Thus the consumers' surplus will be ATURQ, which is smaller than KJLA.

Now we need to establish that there does exist a situation where this will be so, that is, there exist circumstances under which the equilibrium on the new demand curve will lie to the left of J. Consider the demand curve given by (13.3). Clearly, there can be a configuration such that the R-Cournot equilibrium lies vertically below the x-Cournot equilibrium. Let us assume that the amount of foreign exchange with the South is such that this configuration obtains.

Next note that, whenever the x-, and the i-demand curves are linear, the i-Cournot equilibrium will always lie to the left of the x-Cournot equilibrium, as established in the next paragraph. Thus by our assumption, the i-Cournot equilibrium will lie to the left of the R-Cournot equilibrium as well.

Let the x-demand curve be given by:

$$x = a - bp.$$

Then the inverse demand curve would be:

$$p = a/b - x/b.$$

For duopolists 1 and 2, with identical costs of ec per unit, the Cournot equilibrium works out to be:

$$x_i = (a - bec - x_j)/2 \qquad j \neq i,\ i = 1, 2.$$

Thus

$$x_1 = x_2.$$

Thus

$$x_1 = (a - bec)/3 = x_2.$$

The total industry output will be

$$x = 2(a - bec)/3.$$

This will be sold at

$$p = a/3b + 2ec/3.$$

This point is shown by E_1 in Figure 13.8.

When the South takes a loan to buy the good (that is, it is operating on the i-

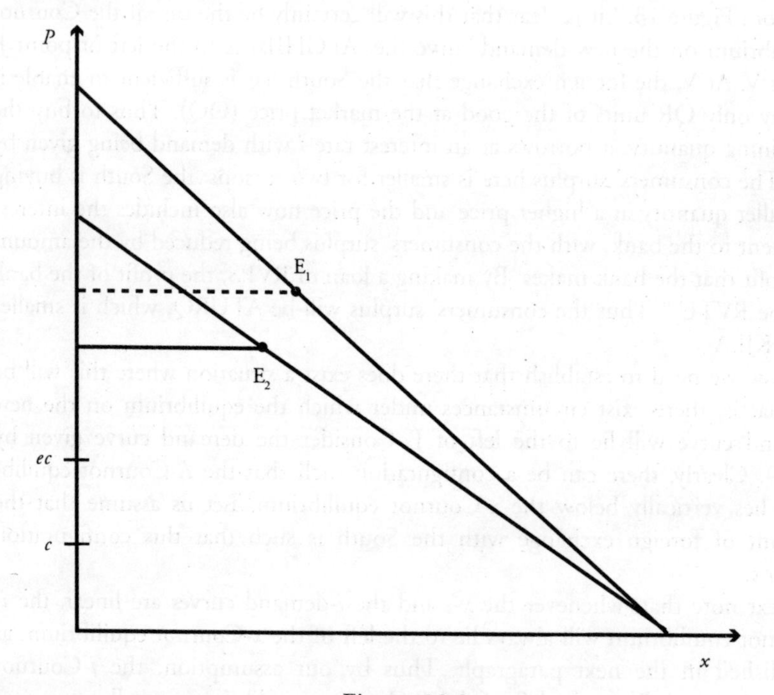

Figure 13.8

demand curve), for each unit of the good, it not only pays the price but also an interest on the loan. Thus the inverse demand curve would now be modified as follows:

$$p + ip = a/b - x/b,$$

or,

$$p = a/(1 + i)b - x/(1 + i)b.$$

The Cournot equilibrium for this curve would be

$$x = 2(a - (1 + i)bec)/3$$

and

$$p = a/3(1 + i)b + 2c/3$$

which is shown as E_2 in Figure 13.8. It is clear from the above that E_2 will always be to the left of E_1.

Now look at Figure 13.7 again. If V is the *i*-Cournot equilibrium, we have to ascertain how this will shift when the demand curve is AFGHIB instead of CB. Theoretically this could shift to any part of the new demand curve, but given that the international lending is done by a profit maximizer, it follows that the equilibrium must lie on the linear stretch GH, corners excluded. If *i* is chosen so that the equilibrium lies on AF, FG, HI, or IB, the lender's profit would be zero and this would be irrational for the lender. It can be demonstrated quite easily that if the Cournot equilibrium remains on the line segment GH, corners excluded, after the demand curve changes from CB to AFGHIB, then the Cournot-equilibrium point must continue to be at V (that is, the *i*-Cournot equilibrium point).

To see this, imagine a situation when the equilibrium on the new demand curve shifts to a point like W. This means that at least once of the duopolists can increase his profits by moving away from V towards W. But W was available even earlier, which was rejected, and V was chosen as the Cournot-equilibrium point. Thus W cannot be an equilibrium point now. Thus V will continue to be the Cournot equilibrium on the new kinked demand curve.

The proof of our proposition is now complete. To summarize, under certain assumptions, we know that the *i*-Cournot equilibrium will lie to the left of the *R*-Cournot equilibrium (the latter being the original equilibrium in the absence of international lending). With international lending, if the equilibrium continues to be at the *i*-Cournot point, it will necessarily lie to the left of the original equilibrium. When international lending is done by a profit-maximizing bank, the final equilibrium must be an *i*-Cournot equilibrium. This situation, as we have seen, results in a smaller consumers' surplus for the South. This is the state under which the welfare of the South is reduced by resorting to international borrowing.

To see how the adverse-effect theorem may be relevant in the kinds of contexts that are used by economists to describe problems of oligopoly, we begin by constructing a numerical example.

Let $x = 10 - p$, $c = 1$, $e = 1$, $R = 12$.

This is a case where the R-Cournot equilibrium lies vertically below the x-Cournot equilibrium. From the method of proof of the adverse-effect theorem used above it will be transparent that the adverse effect indeed occurs in the above case.

It is easy to see that the example is generic. Virtually all the parameters in the above example can be generalized to a class where the adverse effect occurs. Let us, for instance, take the case of R. Note that if $R > 12$, the R-Cournot equilibrium will be to the right of the x-Cournot equilibrium. However, if $R \geq 24$, then the foreign exchange constraint ceases to be binding. Hence, as long as $24 > R \geq 12$, the adverse effect does occur. The other parameters may be generalized in a similar manner.

13.6. Concluding Remarks

Our theoretical analysis leads us to conclude that the commonly held notion that credit boosts donor exports holds true under certain conditions, namely, under market structures of monopoly and perfect competition in the donor (North). If the Northern manufacturing industry is oligopolistic, the results are varied—if international credit is given beyond a certain level, it can lead to a fall in the donor exports. Also, we find that there is an optimum amount of credit that maximizes the profits of the Northern oligopolists.

This highlights the important role of credit as a strategic variable, from the point of view of the North. Insofar as this analysis is valid, the oligopolists can reach an understanding with the lending institutions in order to ensure that the amount of credit given is optimal from the former's point of view. The literature on loan pushing suggests that, in fact, such an understanding has taken place on many occasions.

We have also demonstrated the possibility of the South being worse off as a result of international borrowing. This happens as a consequence of strategic interaction between the agents involved, to wit, the Southern consumers, the Northern manufacturers and the Northern lending bank.

Notes

1. See, for example, Winkler (1929), Hyson and Strout (1968), Gwyne (1983), Taylor (1985), Darity and Horn (1988), and Basu (1991).
2. Bhagwati (1967) is one of the pioneering works on this question; Jepma (1991) contains a review of the theory and practice of aid-tying. For a model of tied-in credit in a monopoly market, see Besley (1988). Besley's model, however, pertains to an indigenous market with domestic credit.
3. See, for instance, Ohlin (1966), Hayer (1971), Wall (1973), Payer (1974), Tendler (1975).
4. The interesting debate in the 1980s on the transfer problem (see, e.g., Chichilnisky 1980, Dixit 1983, and de Meza 1983), which revisits an old controversy, is focused

on the adverse movement (from the recipient's point of view) in the terms of trade as a consequence of receiving aid or credit. But the model there was entirely non-strategic, with all agents engaged in price-taking behaviour. In our model, the interaction is strategic and, as will be evident from the sections that follow, this entails a very different kind of analysis.

5. Although much of the lending during the 1970s was sovereign lending, we make no distinction between the debt contracted by the government in the South or by individual corporations in the South. The distinction is immaterial from the point of view of the model because even if the loan is taken by, say, a corporation in the Southern country, its government can back the loan, thus obliterating any useful distinction between the two agents.
6. This being a partial-equilibrium model, e and R are exogenous, and it is their effect on the demand for the manufactured good that is being discussed. However, in some LDCs, the demand for an imported manufactured good can have important feed-back effects on e and R through trade. For instance, an increase in the quantity of x could lead to an increase in the South's exports and hence in R, at the given exchange rate.
7. More complicated cases can arise if the demand curves intersect more than once. These are easy to analyse using the method used below and, moreover, there will be no occasion in this chapter to consider such complications. It is therefore adequate here to focus on Figure 13.3.
8. Despite the sophistication of this demand function, it leaves out many features of reality. We do not, for instance, go into the nature of the market for this good in period 2. Moreover, there may exist more than one good for the purchase of which hard currency is needed. Hence a limited amount of fungibility would occur here and the ideal way to work out the problem may be by using two budget constraints. While we do not go into this here, we have worked much of this out for ourselves. The main results of this chapter remain unaffected.
9. The opportunity cost of lending is being treated as zero. This causes no loss of generality.

References

Basu, K. 1991. 'The International Debt Problem, Credit Rationing and Loan Pushing: Theory and Experience'. *Princeton Studies in International Finance* 70.

Besley, T. 1988. 'Tied-in Credit with a Monopoly Product Market'. *Economics Letters* 28: 105–8.

Bhagwati, J. 1970. 'The Tying of Aid'. In J. Bhagwati and R. Eckaus, eds. *Foreign Aid.* Harmondsworth: Penguin Books.

Chichilnisky, G. 1980. 'Basic Goods, the Effect of Commodity Transfers and the International Economic Order'. *Journal of Development Economics* 7: 505–19.

Darity, W., Jr and B. Horn. 1988. *The Loan Pushers: The Role of the Commercial Banks in the International Debt Crisis.* Cambridge, Mass.: Ballinger.

De Meza, D. 1983. 'The Transfer Problem in a Many Country World: is it Better to Give than to Receive?'. *Manchester School* 51: 266–75.

Dixit, A. 1983. 'The Multicountry Transfer Problem'. *Economics Letters* 13: 49–53.

Dixit, A. 1984. 'International Trade Policy for Oligopolistc Industries'. *Economic Journal* (Supplement) 94: 1–16.

Gwyne, S.C. 1983. 'Adventures in the Loan Trade'. *Harper's Magazine* (September): 22–6.

Hayter, T. 1971. *Aid as Imperialism.* Harmondsworth: Penguin Books.

Hyson, C.D. and A.M. Strout. 1968. 'Impact of Foreign Aid on US Exports'. *Harvard Business Review* 46: 63–71.

Jepma, C. 1991. '*The Tying of Aid.* Paris: OECD.

Ohlin, G. 1970. 'The Evolution of Aid Doctrine'. In J. Bhagwati and R. Eckaus, eds. *Foreign Aid.* Harmondsworth: Penguin Books.

Payer. C. 1974. *The Dept Trap.* New York: Monthly Review Press.

Pincus, J.A. 1963. 'The Cost of Foreign Aid'. *Review of Economics and Statistics* 45: 360–7.

Rothermund, D. 1981. 'British Foreign Trade Policy in India During the Great Depression, 1929–1939'. *Indian Economic and Social History Review* 18: 349–76.

Taylor, L. 1985. 'The Theory and Practice of Developing Country Debt. An Informal Guide for the Perplexed'. *Journal of Development Planning* 16: 195–277.

Tendler, J. 1975. *Inside Foreign Aid.* Baltimore: Johns Hopkins University Press.

Wall, D. 1973. *The Charity of Nations: the Political Economy of Foreign Aid.* New York: Basic Books.

Winkler, M. 1929. *Investments of US Capital in Latin America.* Boston: World Peace Foundation Pamphlets.

14 Why are so many Goods Priced to End in Nine? And Why this Practice Hurts the Producers

14.1. The Problem

Our local store is offering Women's Leggings for $9.99, Men's Thermal Henleys for $14.99 (these were, by the way, originally available for $19.99), and Classic Fleece Sweatshirts for $24.99. Pizza Hut has been kind and put a flyer in our mailbox with six coupons: two medium two-topping pizzas for $10.99, one medium specialty pizza for $8.99, Family Pairs (whatever that may be) for $12.99, and three other variations for $10.99, $15.99 and $10.99, respectively. When one moves on to more valuable items, the cents vanish but the parade of nine or near-nine endings continues. The latest *Ithaca Times* has an advertisement offering the following seven-day winter getaways to Florida: Orlando $379, Ft. Lauderdale $459, St. Petersberg (not the one with the Hermitage) $489.

The question that this chapter investigates is: Why is this so? Clearly it cannot be that demand conditions and marginal-cost conditions are such that the standard neoclassical equilibrium price almost invariably *turns out* to be such as to end in nine. Providence surely has better things to do.

Confronted with similar pricing dilemmas, economists have resorted to two different lines of explanation. One which appeals to 'psychological illusions' in consumers (see, for example, Monroe 1990); and the other (usually, deductively more sophisticated) 'economic explanation', which somehow explains the phenomenon in question without relenting on 'the assumption of perfect rationality. Thus, for instance, consider the fact of periodic and regular sales or price

From *Economics Letters* 54, 1997, 41–4.
I have benefited from discussions with Abhijit Banerji and Ernst Fehr.

markdowns that shops indulge in. The 'psychological' explanation would resort to arguments like how consumers judge quality by looking at the original price tag on a good and so are more prone to buy an item that is marked down than one which has always been at the same price. We now know, however (see, for example, Varian 1985; Sobel 1984), that the same phenomenon can be given an economic explanation terms of search costs or screening.

We return now to the question that this chapter is concerned with. For brevity, it will be referred to as the phenomenon of 'pricing in the nines'. This is an old phenomenon, usually attributed to custom (Ginzberg 1936). At first sight it seems that the only possible explanation for pricing in the nines is a psychological one. Consumers are busy and their brains have limited storage capacity; so it seems reasonable to suppose that they ignore looking at the last digits of a price. In other words, when they see something priced at D dollars and C cents, they treat this *as if* it cost D dollars (Nagle 1987: 248–9). Given such economizing of brain activity and space use, it makes sense for the producer to make the C as large as possible. Hence, the ubiquitous 99 cents. Consumers, according to this explanation, are systematically deluded.

I show that pricing in the nines has an economic explanation—one in which consumers are perfectly rational and not subject to any illusion. The economic explanation has a very interesting implication. In equilibrium it is the firms that are worse off because of this phenomenon. So, far from hurting the consumers (the equilibrium effect on whom is ambiguous), it is the firms that suffer as a consequence of pricing in the nines. Of course, this being an equilibrium, each firm is individually powerless to change the situation.

14.2. The Solution

For simplicity assume that there are thousands of goods (we may distinguish between the same commodity sold at different locations) and for each good there is a market demand curve and each good is supplied by a monopolist. For good i, let the market demand function be

$$x_i = x_i((D_i, C_i)),$$

where (D_i, C_i) is the price of good i expressed in a somewhat unusual notation. D_i is the number dollars and C_i the number cents. Hence D_i *is* always a non-negative integer and $C_i \in \{0,1, ..., 99\}$. Hence, if the price of good i is 5.75 dollars, we shall say that its price is $(D_i, C_i) = (5,75)$.

It is best to think of us as looking at markets of relatively cheap goods, such that D_i will typically be less than a hundred dollars. If we were looking at more expensive goods, such as automobiles, then I would break up the price between thousands of dollars (T_i) and hundreds of dollars (H_i). Once the reader has understood my argument, translation to the domain of high-priced goods will be transparent and so I confine my attention to D_is and C_is.

It will be shown that under a plausible assumption about how consumers reason (which entails no compromise with the assumption of full rationality), producers will invariably set $C_i = 99$ and they will be worse off as a consequence. The argument of this chapter being sufficiently straightforward, I do not resort to formalism. Also, instead of directly describing the final equilibrium, I describe a possible process leading up to the equilibrium, in the belief that such a method facilitates intuition.

Let us first assume that if consumers behave like textbook consumers, that is, given price (D_i, C_i) they demand $x_i(D_i, C_i)$, then in sector i the monopolist's equilibrium price is (D_i^*, C_i^*). Let $\phi(\cdot)$ be the frequency distribution of the C_i^*s over $\{0,1, ..., 99\}$. That is, for each $C \in \{0, 1, ..., 99\}$, $\phi(C)$ is the number of goods for which the price ends in C cents. More formally,

$$\phi(C) = \#\{i | C_i^* = C\}.$$

It seems reasonable to assume that $\phi(C)$ will be a uniform distribution on $\{0, 1, ..., 99\}$. No such assumption is, however, necessary here.

We will simply assume that consumers, through browsing and the unwitting collection of information, know the distribution $\phi(C)$ that prevails on the market. Being busy, they economize on their brain function as follows. Let the expected value of C be EC. That is, $EC \equiv \Sigma C\phi(C) / \Sigma\phi(C)$. Suppose a consumer has to decide on how many units of good i to purchase. She will, it is being assumed here, not waste effort looking at C_i^*. She simply looks at D_i^* and assumes that C_i^* is the expected value, that is, EC. On average, she will be right and, given some cost to the use of the brain, this could be a perfectly rational way of thinking.

We could go a step further and assume that consumer expectation of C is conditional upon the observed value of D. This will leave the analysis unchanged. Note that I am accepting the view that consumers process prices from left to right but, unlike in the standard marketing literature, I am assuming that what consumers expect for the last digits of a price is fully rational.

If consumers behave as described above, and good i is priced at (D_i^*, C_i^*), then the demand for good i will be $x_i((D_i^*, EC))$. Assuming that there are lots of firms and the effect on EC of a single firm changing its C_i is negligible, it follows that for each firm the optimal C_i to choose is 99.

Hence, each firm i chooses price equal to $(\hat{D}_i, 99)$ where $(\hat{D}_i, 99)$ gives the firm at least as much profit as $(D_i, 99)$, for all non-negative integers D_i. Hence, the final equilibrium-price profile that prevails in the market is $\{(\hat{D}_i, 99)\}_{i=1,2,...}$

Observe that consumers being rational, the demand for good i in equilibrium is given by $x_i((\hat{D}_i, 99))$. There is no psychological illusion. When buying good i, consumers still look only at $\hat{D}_i$ and simply presume that $\hat{C}_i$ takes its expected value, 99. And indeed $\hat{C}_i$ is 99.

But note that barring some coincidental cases these firms do worse than firms in textbook models. To see this note that if a single firm, i, deviates from $(\hat{D}_i, 99)$ and prices its product at $(\hat{D}_i, C_i)$ where $C_i < 99$, then demand will still remain at $x_i(\hat{D}_i, 99)$ since consumers assume $C_i = EC$. If, on the other hand, the firm changes the price to $(\hat{D}_i - 1, 99)$, then demand changes to $x_i(\hat{D}_i - 1, 99)$. And this, by the definition of $\hat{D}_i$, we know is not worthwhile. In other words, in equilibrium, each firm, effectively, faces a step-wise demand function which is to the left of the real demand function $x_i(D_i, C_i)$. Hence, it generically does worse than it would if it had faced the real demand function.

It is worth noting that, in the final equilibrium, the expectation of the consumer that a good which is priced at D dollars and some cents is actually priced at D dollars and 99 cents turns out to be exactly right.

14.3. Conclusion

This chapter does not prove that the psychological explanation is wrong. It merely provides an economic explanation for a phenomenon for which this is not the first explanation that springs to mind and, indeed, the 'evidence' of psychological illusion seems to be all too transparent. It is therefore interesting to see that what seems so transparent is not *necessarily* true and that there is a competing explanation which does not have to sacrifice the assumption of perfect rationality on the part of the consumer. To check whether psychological explanations play a role in industrial strategy and pricing, one would have to go beyond the phenomenon of pricing in the nines and devise subtle empirical tests and perhaps even do experiments. Such tests can only enrich our models of industrial organization and pricing.

References

Ginzberg, E. 1936. 'Customary Prices'. *American Economic Review* 26: 296.

Monroe, K.B. 1990. *Pricing: Making Profitable Decisions*, 2nd edn. New York: McGraw-Hill.

Nagle, T.T. 1987. *The Strategy and Tactics of Pricing.* Englewood Cliffs, NJ: Prentice Hall.

Sobel, J. 1984. 'The Timing of Sales'. *Review of Economic Studies* 51: 353–68.

Varian, H. 1985. 'Price Discrimination and Social Welfare'. *American Economic Review* 75: 870–5.

PART IV
Governments, Games, and the Law

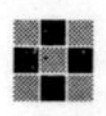

15 Notes on Bribery and the Control of Corruption

with Sudipto Bhattacharya and Ajit Mishra

15.1. The Problem

A person, Z, is considering committing a crime or an act of corruption, like evading income tax or accepting a bribe. Let the benefit to him from this act be *B* units, the probability of getting caught be *p*, and the penalty be *f*. The problem confronting Z is allegedly a simple decision problem in which he has to only calculate the expected cost associated with the corrupt act (in this case it is *pf*) and commit the crime if it is less than the booty (in this case *B*). This is the standard model of crime and corruption (Becker 1968). One can complicate the description by bringing in utility functions, non-neutral attitudes to risk, and wealth effects—but we want to argue here that there is a more fundamental problem with this model and its many variants that have appeared in the literature.

This is especially true in the case of bribery. Suppose that in the above model the crime in question involves person Z *taking* a bribe of amount *B*, and that Z is caught after the crime by policeman 1. In the standard model he then has to pay a penalty of *f* which goes to the government. But clearly we should allow for the fact that he may try to bribe policeman 1. Especially since the original crime was supposed to involve bribery, there is no reason now to assume that bribery is not possible.

If we allow for bribery and policeman 1 is treated as a rational homo economicus, then what we have confronting us now is a standard bargaining problem involving Z and 1. If they fail to reach an agreement (about the

From *Journal of Public Economics* 48, 1992: 349–59.

We are grateful to T.C.A. Anant and Debraj Ray for comments and suggestions. The paper also benefited from a presentation at Penn State University.

appropriate size of the bribe), then we shall suppose that Z will have to pay the penalty f which will be handed over to the government and 1 will get nothing. Hence, the official penalty plays the role of determining where the threat-point (or 'fall-back utility') of the two agents is. It is worth noting straight away that even if no one pays penalties in the society, the size of the penalty can play a role in crime control because it can influence the equilibrium level of the bribe.

In the above bargaining problem we shall determine the outcome by applying the standard Nash solution. But the application of the Nash solution is not as easy as may appear at first sight.

To explain the difficulty let us assume that in this society, bribe-*giving* is not considered a crime; only bribe-*taking* is. Nothing essential is lost by this assumption of asymmetry as is demonstrated later in note 4.

Suppose now that Z gives policeman 1 a bribe of B_1. To compute the Nash solution we need to know how much 1 benefits from this. The problem is complicated by the fact that after 1 takes the bribe, he in turn can be caught by policeman 2. Hence, although he gets B_1 to start with, in the end his expectation is lower. The chain continues and our modelling will depend on whether we think of this as a finite or an infinite chain. Empirically, bribes moving up a hierarchy are well known (see, for example, Wade 1988); our aim here is to give this theoretical structure.[1]

We begin by considering the *infinite*-chain case. To solve the Nash bargaining problem at any stage we need, in some sense, to know what will happen in all future stages. An important aim of this chapter is to formalize the above problem in a way that allows us to pose the infinite regress problem in a manageable way. After doing that, we consider some variants of the model and discuss how the penalty size and the probability of detecting corruption can be used to control corruption in a society.

15.2. Nash Bargaining and Equilibrium Bribe

In developing a formal structure it is useful to begin by restating rigorously some of the remarks of the previous section.

If a person is caught taking a bribe of B units, he is expected to pay a penalty or fine of $f(B)$. So f is the *penalty function* specified by the country's law.

The probability of getting caught after taking a bribe is p. For simplicity we assume that p does not depend on B. The real number p and the function f are both controlled by the government and are exogenous to our model.[2]

Next we define the central feature of our model—the *bribe function*, ϕ. If a person is caught having taken a bribe B, he can get away by paying a bribe of $\phi(B)$. The bribe function is endogenous. It is determined by Nash bargaining and by taking into account the infinite regress problem discussed in Section 1. Fortunately the infinite regress can be captured by a simple one-shot recursive procedure.

We assume that if a bribe-taker is caught, he indulges in Nash bargaining with his captor in deciding on the bribe. If the bargain fails, the bribe-taker has to pay the penalty (which goes to the government) and the captor gets nothing. In working out the bargaining solution it has to be kept in mind that the captor, should he take a bribe, risks being caught by another person.

Keeping this in mind and using Nash bargaining we can define an equilibrium bribe function as follows.

The function ϕ is an *equilibrium* if and only if, for all $B \geqq 0$,

$$\underset{B'}{\text{argmax}}\ [f(B) - B'][B' - p\phi(B')] = \phi(B).$$

To understand the equilibrium definition fully, it may be useful to write the above equation a little more elaborately as follows:

$$\underset{B'}{\text{argmax}}\ [(B - B') - (B - f(B))][B' - p\phi(B')] = \phi(B).$$

Suppose person n has been caught by person $n + 1$ for having taken a bribe of B. Now n is trying to give a bribe to $n + 1$. If he can get away by giving a bribe of B', then his net gain is $[(B - B') - (B - f(B))]$. This is so because if the bargain fails, n falls to n's threat level which is $B - f(B)$. Now look at it from $n + 1$'s point of view. If he accepts a bribe of B', his net gain is $B' - p\phi(B')$. It must be remembered that on his way home with a bribe of B' there is a probability p he will be caught and he will then in turn have to pay a bribe of $\phi(B')$. Since we are interested in the Nash bargaining solution we find out the equilibrium value of B' by maximizing the multiplication of the net gains.

Since the equilibrium B' thus derived is the bribe that n has to give for having taken a bribe of B, B' must be equal to $\phi(B)$.

In the completely general case the existence of equilibrium is difficult to prove, let alone characterize. Also, there is a problem of interpreting the Nash solution. As is well known, the standard Nash solution was developed for the case where the feasible set is convex (see, for example, Friedman 1986). Although some of the other bargaining solutions, like that of Kalai and Sinorodinsky (1975), can be adapted for the non-convex case, the same is not true of the Nash solution (Anant, et al. 1990).

Here we use a simplifying assumption that ensures the relevant convexity, guarantees existence, and allows us to characterize the equilibrium. From now on we focus on a linear penalty function:

$$f(B) = FB, \text{ where } F > 0. \tag{15.1}$$

In this case the equilibrium bribe function is easily deduced to be

$$\phi(B) = FB/2. \tag{15.2}$$

Hence, the amount of the bribe that has to be paid is half the amount of the fine.

Let us check, in terms of the exogenous variables p and F, as to how corruption may be controlled in this society.

It will not be worthwhile for a person to take a bribe of B if his net expected gain from this is non-positive. Clearly his net expected gain, given (15.1) and

(15.2), is given by $X \equiv (1-p)\,B + p\,[B - (FB/2)]$. If bribing is to be stopped, X has to be less than or equal to zero for all B. This will be so if and only if

$$(15.3) \qquad p^F \geqq 2$$

What is interesting is to compare this with a society in which if you are caught for taking a bribe, B, you have to pay the fine FB. This is the standard model. In the standard model taking a bribe is not worthwhile if and only if

$$(15.4) \qquad (1-p)\,B + p\,[B - FB] \leqq 0 \text{ or } pF \geqq 1.$$

Comparing (15.3) and (15.4) it is clear that, once we make allowance for using bribery to get away from having to pay the penalty for an initial crime, the problem of crime control is much more difficult (pF has to be $\geqq 2$) than is supposed on the basis of the standard model (where pF has to be $\geqq 1$).

Observe also that although in our model (suppose $pF < 2$) the penalty is never paid, the penalty nevertheless is an instrument that can be used to curb corruption. It is a less efficient instrument than originally supposed, but an instrument all the same. So the wisdom occasionally expressed in popular fora, namely that if a penalty is always evaded through bribery then we may as well not have the penalty, is wrong. The penalty influences the equilibrium bribe and can therefore be an indirect instrument of control. This is, in a sense, a similar idea to the one developed in Huberman and Kahn (1988), where agents write contracts knowing that these will be renegotiated away later on. This is because the initial contract influences the threat-point of the renegotiation bargain.

15.3. Elite Forces and Finiteness

F and p are not the only variables a government uses to control crime and corruption. Creating a special elite civil service that is highly paid and trained to block (at least petty) corruption has been tried by most nations. To see the impact of this, let us incorporate this into our model of Section 15.2. There are two ways of doing this.

15.3.1. The Hierarchical Elite

First, we shall suppose that if a person Z is caught for an initial crime, he can bribe policeman 1 to get away, policeman 1 can bribe policeman 2, etc.; but the chain is not endless. After the nth round of bribery the police officer one encounters belongs to the elite civil service and he (yes, let us assume) is incorruptible. In other words, we wish to consider a finite model analogue of the previous model. In studying this we shall retain the linearity assumption (15.1).

We have to solve the problem by backward induction. Consider the nth round. Person i who had taken a bribe of B has been caught by person j. Now i is trying to bribe j. In working out the Nash bargaining solution, we have to keep in mind that if j accepts a bribe of B' and is then himself caught, he will have to pay a penalty of FB'. There is no further getting away. Hence the Nash bargaining solution between i and j is given by:

$$\underset{B'}{\text{argmax}}\ [FB - B'][B' - pFB'] = \phi_n(B), \tag{15.5}$$

where $\phi_n(B)$ is the bribe a person has to give in the nth round if he is caught having taken a bribe of B. It is easy to check that, if $pF < 1$, then

$$\phi_n(B) = FB/2, \tag{15.6}$$

which is exactly the same as (15.2). This being so, for all earlier stages, t, the bribe function will be exactly the same, namely $\phi_t(B) = FB/2$.

In other words having an incorruptible elite service blocks corruption after the nth round but, surprisingly, leaves the structure of bribery unchanged in the earlier stages.

It is, however, interesting to note that in this case the conditions for the control of corruption are not the same as those following (15.2), although (15.6) has the same form as (15.2). To see this, note that (15.6) is valid if $pF < 1$. This is because $pF > 1$ implies that $B' - pFB' < 0$. From (15.5) it is clear that the nth bribe-taker (that is, the last bribe taker) would in such a situation make a loss by taking a bribe B', no matter what value B' takes. Hence the nth person would not take a bribe if $pF \geqq 1$.

Suppose now $pF \geqq 1$, and consider the $(n - 1)$th policeman. Since the nth policeman will not take a bribe, it is *as if* the nth policeman belongs to the incorruptible elite service. So our $(n - 1)$th policeman's decision problem is identical to the nth Policeman's decision problem discussed in the above two paragraphs. But in that case we know that if $pF \geqq 1$, the policeman will not take a bribe. So the $(n - 1)$th policeman will not take a bribe. It follows, by backward induction, that if $pF \geqq 1$, corruption will not occur.[3]

15.3.2. The Mingling Elite

It is possible that the incorruptible policemen (elites) are uniformly distributed across different layers ($n = 1,2,\ldots$), unlike the previous case where one particular layer consisted only of elites. Assuming the fraction of incorruptible policeman in each layer to be the same (say equal to d), d can be viewed as the probability that a corruptible officer (or person Z) assigns to his superior officer (the one who apprehends him) being of the incorruptible type.

Following our earlier method it is obvious that the equilibrium bribe, $\phi(B)$, is worked out from the following:

$$\underset{B'}{\text{argmax}}\ [FB - B'][B' - p\{(1 - d)\phi(B') + dFB'\}] \equiv \phi(B).$$

Differentiating the left-hand expression with respect to B', setting it equal to zero, and solving it, gives us $B' = FB/2$ or $\phi(B) = FB/2$.

To deter corruption the government will now have to set

$$p(1 + d)F \geqq 2. \tag{15.7}$$

This is easily checked. To select an optimal control mechanism the government will have to figure out the costs of raising p, d, and F, then minimize this sum subject to (15.7). It is interesting to note that if $d = 0$, (7) becomes $pF \geqq 2$, which is our model of Section 15.2, and if $d = 1$, we get $pF \geqq 1$, which is the conventional view.

15.3.3. The Finite Case with no Elite

To complete our line of enquiry we ought to discuss a case where the hierarchy is finite, as in Subsection 15.3.1, but there is no elite force at the top of the hierarchy. The chain of auditors and super-auditors stops at the nth level, and at the nth level a bribe can be taken with no fear of being apprehended. In this case if the $(n - 1)$th policeman is caught by the nth policeman for having taken a bribe of size B, the bribe he has to give in order to get away, $\phi_n(B)$, is clearly given by

$$\underset{B'}{\operatorname{argmax}}\, [FB - B'][B'] = \phi_n(B).$$

It follows that $\phi_n(B) = FB/2$. Moreover, for all earlier stages the same is true. That is, $\phi_t(B) = FB/2$, for all $t = 1, ..., n$.

15.4. Corruption and Chain Arrests

We have thus far worked with the assumption that when a person is apprehended for some act of corruption, the chain of corruption preceding him is left unearthed. Another possible assumption is to go to the other polar end and suppose that if the kth policeman is caught for having taken a bribe, then *all* the earlier bribe-takers in the chain are caught. It is difficult to decide a priori as to which is a better assumption. If a person is caught red-handed taking a bribe, it may be realistic to assume that the entire chain of corrupt acts, starting from person Z, can be unearthed. Therefore, in the absence of empirical evidence it is worthwhile considering the case of 'chain arrests' as well.

The infinite case (that is, the counterpart of the model of Section 15.2) is straightforward. We shall throughout this section make one assumption for simplicity. Suppose the kth policeman unearths a chain of bribe taking from person Z, through policeman 1 to policeman $(k - 1)$. We shall assume that each arrested person bargains independently with his arrester, namely policeman k. In the infinite case the probability of person Z being caught for having taken a bribe, B, by policeman 1 is p. Suppose he bribes him and 'gets away'. The probability of policeman 1 being caught by policeman 2 and therefore of Z being caught by policeman 2 (remember arrests occur in chains now) is p. Therefore, when Z takes a bribe of B, the probability of being caught by policeman 2 is p^2, and, by an extension of the same argument, the probability of being caught by policeman t is p^t. It follows that the bribe he has to give policeman 1, $\phi(B)$, is given by

$$\underset{B'}{\operatorname{argmax}} \left[FB - \left(\frac{1}{1-p}\right)B'\right]\left[B' - p\left(\frac{1}{1-p}\right)\phi(B')\right] = \phi(B),$$

since $1 + p + p^2 + \ldots = 1/(1 - p)$. Hence, $\phi(B) = (1 - p)\, FB/2$.

It follows that the expected aggregate bribe that Z has to pay is

$$\frac{(1-p)FB}{2}(p + p^2 + \ldots) = p\frac{FB}{2}.$$

As before, corruption is curbed if $pF > 2$.

The finite case (that is, the counterpart of the model of Subsection 15.3.3) is, however, much more interesting. Let us suppose that the nth stage is the last one and there is no incorruptible elite force involved. Let B be the amount of bribe taken by the ith policeman and B_0 the bribe taken by Z. Consider the nth round. Thus policeman n has arrested all the bribe-takers. Let $\phi_{ni}(B)$ be the bribe that person i has to pay to policeman n when arrested by n. We follow the convention that if $i = 0$, then the reference is to person Z. Since getting away at the last stage means complete acquittal, it follows that

$$\phi_{ni}(B) = \underset{B'}{\text{argmax}}\, [FB - B'][B'] = \frac{FB}{2'} i = 0, \ldots, n-1.$$

Now consider the penultimate round. Thus policeman $(n - 1)$ has made a chain arrest of 0, 1, ..., $(n - 2)$ for having taken bribes (remember 0 here refers to Z). Clearly if i has taken a bribe of B he has to pay policeman $(n - 1)$ a bribe as given below

$$\phi_{(n-1)i}(B) = \underset{B'}{\text{argmax}}\left[FB - B' - p\frac{FB}{2}\right]\left[B' - p\,\frac{FB'}{2}\right]$$

$$= \left(1 - \frac{p}{2}\right)\frac{FB}{2}, \; i = 0, \ldots, n-2.$$

Proceeding similarly, it can be shown through some tedious but simple algebra that if Z is caught by 1 for having taken a bribe of B, he has to pay a bribe of

$$\left(1 - \frac{p}{2} - \frac{p^2}{4} - \ldots - \frac{p^{n-1}}{2^{n-1}}\right)\frac{FB}{2}.$$

Hence the *total* expected bribe payment by Z would be given by

$$p\left(1 - \frac{p}{2} - \ldots - \frac{p^{n-1}}{2^{n-1}}\right)\frac{FB}{2} + p^2\left(1 - \frac{p}{2} - \ldots - \frac{p^{n-2}}{2^{n-1}}\right)\frac{FB}{2} + \cdots + p^n\,\frac{FB}{2} \tag{15.8}$$

$$= p\,\frac{FB}{2}\left(1 + \frac{p}{2} + \ldots + \frac{p^{n-1}}{2^{n-1}}\right).$$

For large n—more precisely, as n goes to infinity—(15.8) reduces to $pFB/(2 - p)$. Hence, Z would not commit the initial crime of B if and only if

$$B \leqq \frac{pFB}{2-p}.$$

In other words, corruption is curbed if

$$2 \leq p(F + 1).^4 \tag{15.9}$$

Expression (15.9) is interesting because, unlike in all the other cases considered in this chapter, p and F are shown here to play asymmetrical roles in the control of corruption. There is a widespread view that since increasing the probability of detecting corruption, namely p, is usually an expensive affair entailing an expansion of the police force, whereas raising the penalty, F, can be achieved relatively costlessly, by the stroke of a magisterial pen, it is better to control corruption by a greater reliance on F. Condition (15.9) coaxes us gently in the other direction since it shows that raising p a little has a more powerful thwarting effect on corruption than is suggested by the conventional models as captured by (15.4), or even (15.3).

15.5. Concluding Remarks

There are several directions that can be pursued from here. An obvious venture is to endogenize p and d. Recall that d is the probability that, if one is caught by a policeman, the policeman will turn out to be incorruptible. Moreover, an incorruptible policeman or auditor in our model is someone who does not accept bribes, no matter what the net benefits. It could, however, be argued that although an incorruptible person never takes a bribe, the fraction of the police and auditor population that chooses to be incorruptible depends on how 'expensive' it is to be incorruptible (not on a case-by-case basis, but in equilibrium). This would endogenize d and raise the possibility of multiple equilibria since it may be more expensive to be incorruptible if more people are corrupt.[5]

Turning to p, note that while it is the size of the elite force, within the police force or tax collectors that affects d, p is directly related to the size of the police force or tax collectors, C. To the extent that raising C is costly, a government may have to settle for a lower p than is technically possible. It is possible to derive more specific expressions by attributing a specific social welfare function to the government and filling in the positive features of the model.

Our model suggests another route for endogenizing p which does not require C to be endogenous. Note that the probability of detecting corruption depends not just on the size of the police force but also on the effort it exerts. Increased effort diminishes a person's utility through the reduction of leisure, but in our model, because of the presence of bribery, there is an offsetting aspect to this. Increased effort at catching corrupt people enhances one's 'bribery income'. Since an enhanced effort raises p, a fruitful line of enquiry would be to endogenize p by bringing the auditor's utility function, with leisure and income as arguments, into the analysis.[6]

Another matter worthy of future investigation is the use of rewards to auditors and policemen for reporting corruption. This can be especially interesting in the context of our model because while several variants show up the aggregate expected bribe by Z to be $pFB/2$, the underlying structure changes in ways such that different reward schemes can thwart corruption altogether. Thus, for instance, in some cases $pFB/2$ is obtained by summing over small expected bribes over long chains. Presumably in such cases a 'small' reward scheme can be incentive enough for an auditor to turn in a corrupt person because the bribe he can hope for, being just one person in the chain, will be relatively small.

In closing we wish to emphasize that the objective of this chapter was simply to highlight a problem in the existing literature and to suggest a simple method for overcoming it. This method can be put to several uses. In this section we have tried to do no more than indicate some of these.

NOTES

1. Cadot (1987) has also discussed bribery in a model with a hierarchical administration. But his model is different and, in particular, is not concerned with the problem of recursion which is central to our analysis. See also, Rose-Ackerman (1975) for a model in which the government uses an intermediate agency to perform certain economic tasks.
2. There is an incentive aspect to bribe, stemming from the fact that, after apprehending a criminal, a policeman is entitled to a bribe income. We would therefore expect the probability of catching a criminal to depend positively on the size of the bribe. In a larger setting, bribes could affect the *nature* of economic equilibrium. For instance, in a system of queuing for government licences or quotas, bribes could be a mechanism for moving up the queue (for related work, see Lui 1985).
3. As always, the backward-induction argument is predicated upon a tenuous information structure. It presupposes that policeman 1 knows that policeman 2 knows that ... that policeman $(n - 1)$ knows that policeman n is incorruptible. Small violations of this information assumption could alter the results as we know from standard works in game theory (e.g. Kreps et al. 1982).
4. We have throughout assumed that only bribe-*taking* is a crime. It is easy to extend our model to the case where *giving* a bribe is also considered a crime. Remaining within the linear framework, suppose the penalty for *giving* a bribe of B is GB. Given this assumption, it is possible to check by the method of Nash bargaining and recursion, used above, that if Z is caught for having taken a bribe of B, he will have to pay policemen 1 a bribe of $FB/(2 + pG)$ in order to get away. And it is easily checked that the *total* expected bribe payment (i.e., once for taking and several times for giving) would be once again $pFB/2$.
5. The likelihood of multiple equilibria in models of corruption has been noted in the literature (e.g. Lui 1986; Cadot 1987).
6. Mishra (1991) has worked out a model with endogenous p and d in the case of $n = 2$.

REFERENCES

Anant, T.C.A., K. Basu, and B. Mukherji. 1990. 'Bargaining without Convexity: Generalizing the Kalai–Smorodinsky Solution'. *Economics Letters* 33: 115–19.

Becker, G. 1968. 'Crime and Punishment: An Economics Approach'. *Journal of Political Economy* 76: 169–217.

Cadot, O. 1987. 'Corruption as a Gamble'. *Journal of Public Economics* 33: 223–44.

Friedman, J. 1986. *Game Theory with Applications to Economics*. Oxford: Oxford University Press.

Huberman, G. and C. Kahn. 1988. 'Limited Contract Enforcement and Strategic Renegotiation'. *American Economic Review* 78: 471–85.

Kalai, E. and M. Smorodinsky. 1975. 'Other Solutions to Nash's Bargaining Problem'. *Econometrica* 43: 513–18.

Kreps, D., P. Milgrom, J. Roberts and R. Wilson. 1982. 'Rational Cooperation in the Finitely-repeated Prisoner's Dilemma'. *Journal of Economic Theory* 27: 245–52.

Lui, F.T. 1985. 'An Equilibrium Queueing Model of Bribery'. *Journal of Political Economy* 93: 760–81.

——. 1986. 'A Dynamic Model of Corruption Deterrence'. *Journal of Public Economics* 31: 1–22.

Mishra, A. 1991. 'Does Higher Punishment Imply Less Corruption?'. Delhi School of Economics. Mimeo.

Rose-Ackerman, S. 1975. 'The Economics of Corruption'. *Journal of Public Economics* 4: 187–203.

Wade, R. 1988. *Village Republics: Economic Conditions for Collective Action in South Asia*. Cambridge University Press.

16 A Model of Monopoly with Strategic Government Intervention

with T.C.A. Anant and Badal Mukherji

16.1. Introduction

Government intervention is traditionally characterized as a non-strategic activity. A more realistic approach to model government is to treat it as an entity that may have its own objective function but in strategic terms is no different from other agents, especially large ones like monopoly houses and multinationals. In analysing its interaction with large firms it may be best to treat the government as just another agent in a strategic environment and analyse the properties of Nash equilibria and some of its refinements. Indeed, instances where a multinational far exceeds its host country in economic power are numerous, especially in the Third World. Kindleberger (1984) and Casson (1987) are two examples from a vast literature that views multinational and government interaction in bilateral terms.

The central aim of the chapter is (i) to characterize the Nash equilibrium of the government–firm game, (ii) to demonstrate that 'strategic' inefficiency (explained in this section and formally defined later) is pervasive in these models whenever government uses ad valorem taxes; and (iii) if the government's choice between ad valorem tax and specific or per-unit tax is endogenized, then in a perfect equilibrium the government *will,* in fact, choose the ad valorem system. Observe that (iii) reinforces the significance of (ii).

In Section 16.2 the model is outlined and the existence of equilibrium is established. We consider an industrial situation in which there is a firm that

From *Journal of Public Economics* 57, 1995: 25–43.

We wish to thank Avinash Dixit, Heraklis Polemarchakis, Debraj Ray, and Nirvikar Singh for comments and suggestions. We have also benefited from seminars at the Indian Statistical Institute, Delhi, the Delhi School of Economics, Princeton University, and the University of Stockholm.

sets price in order to maximize profits, and a government that sets ad valorem tax rates in order to maximize revenue collection. Much of the existing public economics literature analyses the consequences of changing tax rates (see, for example, Stern 1987) and could be thought of as a step towards the Stackelberg-type analysis of government–firm interaction. We focus on the Nash equilibrium of such a game.

The revenue-maximization assumption of the government deserves comment. First, our chapter shows how to model government–firm interactions and, therefore, could act as a guide to other models where the government's maximand is different. Second, and more importantly, we believe that governments have different objectives when they undertake different activities and when it comes to taxation policy a typical government's objective does turn out to be revenue-maximization. This could be because, on the expenditure side, there are large pre-commitments such as defence, judiciary, and administrative expenses. Hence, when it comes to taxation, the objective becomes a simple one of trying to keep the budgetary deficit low; that is, to maximize revenue collection. Recently, von Furstenberg et al. (1986) find that the sequence 'spend now–tax later' has much more empirical validity than the conventionally accepted pair of 'spend and tax jointly'. Finally, our assumption fits well standard models of bureaucratic behaviour (see, for example, Niskanen 1971; McGuire et al. 1979).

In Section 16.3, a geometric technique of analysis is developed and our central result—that of strategic inefficiency—is established geometrically in the linear special case. While results of inefficiency are well known in the public economics literature, what we establish is different. In a subgame perfect equilibrium of a two-period model we show that a firm may commit itself to a 'high' cost function (by, for example, signing high-wage contracts with workers, buying outdated technology, or having a cumbersome management structure), which is very distinct from choosing a suboptimal point on the given cost curve. We refer to this as 'strategic inefficiency'.[1]

The strategic-inefficiency result is extremely robust. Section 16.4 generalizes the result by dropping not only linearity but also differentiability, and proves a formal mathematical theorem.

In Sections 16.5 and 16.6 we examine the robustness of these results to alternative specifications. In Section 16.3 we assume that in period 1 the firm chooses its technology (cost function), and in period 2 the firm and government choose price and tax rate simultaneously. In Section 16.5 we consider a three-period model in which the sequence of moves is: cost function first, tax rates second, and price third.

In Section 16.6 the government is allowed (i) to choose between specific or per-unit tax rates and ad valorem taxes and (ii) to have a social-welfare-type objective function. It is interesting to note that the results of the chapter remain valid even under these modifications.

Our model should be treated as illustrative but we believe that it provides a rich base for further theorizing about the consequences of strategic government intervention. We feel that, unlike in many advanced industrialized countries, the main strategic interaction of a firm, especially a multinational giant in a developing economy, is not vis-à-vis other firms but vis-à-vis the government.

16.2. Definitions and the Basic Framework

Let $q(\cdot)$ be the *demand function.* Thus if w is the price faced by consumers, $q(w)$ denotes the aggregate demand for the good. We denote $q(0)$ by Q. We assume that (i) the demand function is continuous and $q(\cdot)$ is zero if the price exceeds a given number P and (ii) the demand curve is downward sloping.

Denoting the *inverse demand function* by $r(\cdot)$ we shall assume that (iii) the total revenue function $x \cdot r(x)$ is concave in x.

The cost function, $c(\cdot)$ is (iv) continuous, convex, and has zero fixed cost and (v) there exists a positive real number, b, such that $c(x) \geq bx$, for all x.[2]

In our model, there is a firm (a monopoly) which chooses the *producer price,* p, to maximize profit, π, and the government chooses an ad valorem tax rate, t, to maximize tax revenue. R. (We later consider modifications of this). Given p and t, the *consumer price* is given by $(1 + t)p$ and the aggregate demand for the good is given by $q((1 + t)p$.

The firm's profit, π, is given by

$$\pi(p, t) = pq((1 + t)p) - c(q((1 + t)p)). \tag{16.1}$$

The government's revenue, R, is given by

$$R(p, t) = tpq((1 + t)p). \tag{16.2}$$

We define (p^*, q^*) to be an equilibrium if

$$\pi(p^*, t^*) \geq \pi(p, t^*) \quad \text{for all } p$$

and

$$R(p^*, t^*) \geq R(p^*, t), \quad \text{for all } t. \tag{16.3}$$

Theorem 16.1: In the above model, given assumptions (i)–(v) an equilibrium exists.

A proof of this appears in Appendix 16.1.

16.3. A Geometric Characterization and Strategic Inefficiency

Let us show a strikingly simple characterization of the above equilibrium in the linear special case. Assume that the demand function is linear and the marginal cost curve is horizontal. These are illustrated in Figure 16.1.

Let PA denote the marginal revenue curve. Through point Q, draw a ray in the north-west direction. Let B be the point where the ray cuts the vertical axis, D the point where it cuts the marginal cost curve, and E where it cuts the marginal revenue. Now, consider the ray that has the property, $BE = ED$. Let BQ be such a ray. From E draw a horizontal (resp. vertical) line and mark its

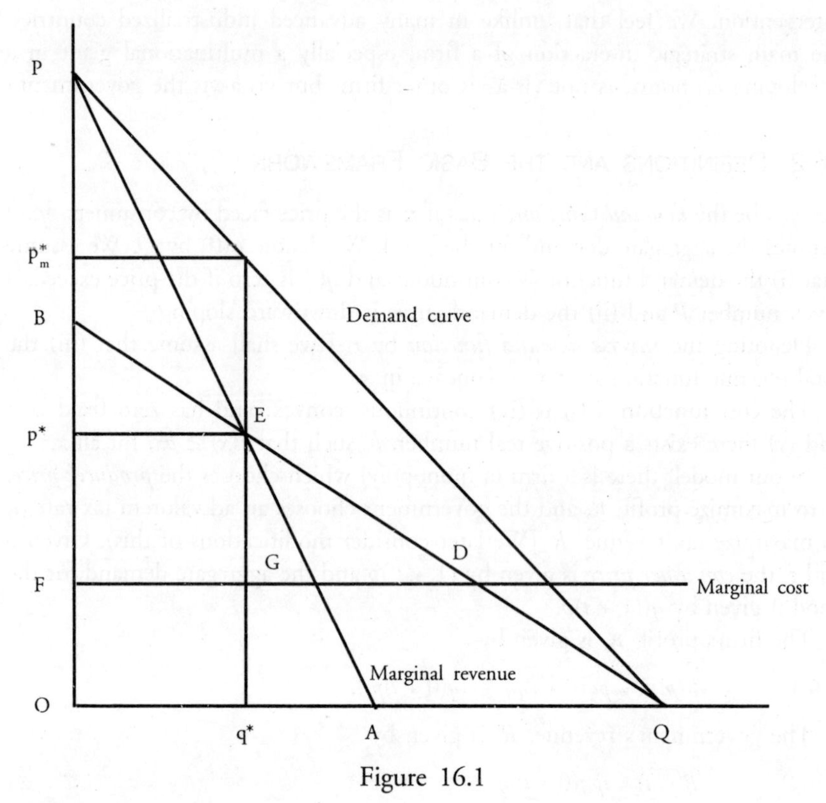

Figure 16.1

intersection with the axis as p^* (resp. q^*), and let $t^* = OP/OB - 1$ and P^*_m be the corresponding market price.

ASSERTION 16.1: (p^*, t^*) is a Nash equilibrium in the one-period model of Section 16.2

We shall refer to this method of spotting the Nash equilibrium in the linear case as the bisection rule (since it involves bisecting the line *BD*) and the ray *BQ*, which has the property that *BE* = *ED*, will be called the equilibrium ray.

PROOF OF ASSERTION: First check how the firm chooses p, given t. In Figure 16.2, let *PQ* be the demand curve and t the ad valorem tax rate. Let *TQ* be a line such that its height is

$$\left(\frac{p}{1+t}\right)$$

of the height of the demand curve. Hence *TQ* shows for each q the producer price. Treating this as a demand curve, the equilibrium is determined by the intersection of the 'marginal-revenue' curve (line *TZ* in Figure 16.2), with the marginal-cost curve (line *CC*). In what follows we shall refer to a curve like *TQ* as the t-demand curve. It will shift as the tax rate, t, changes.

Now return to Figure 16.1. Let the tax rate be such that the t-demand curve is *BQ*. Since *BE* = *ED*, hence *FG* = GD. Hence, the marginal revenue for *BQ*

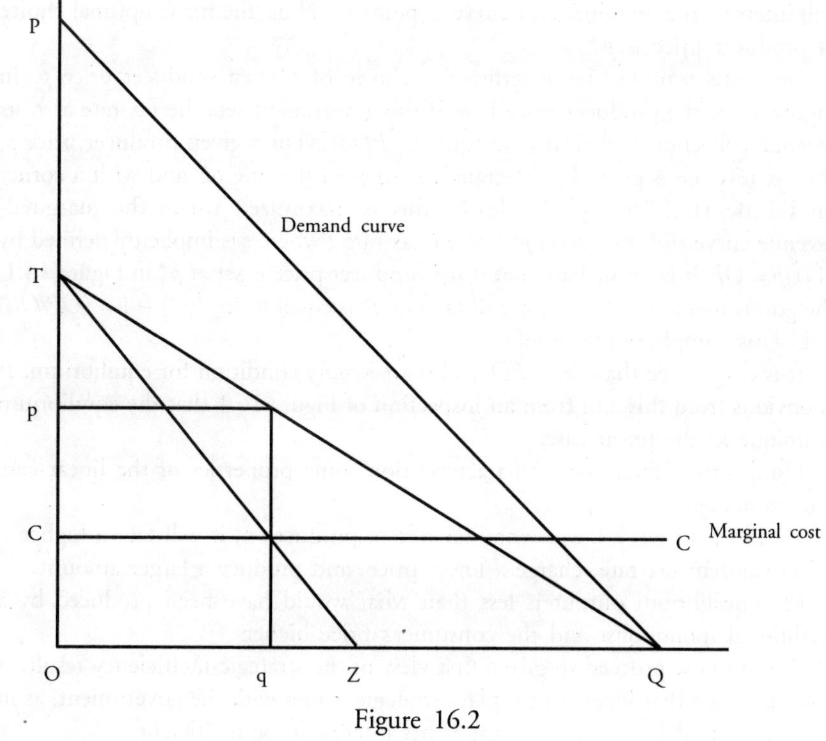

Figure 16.2

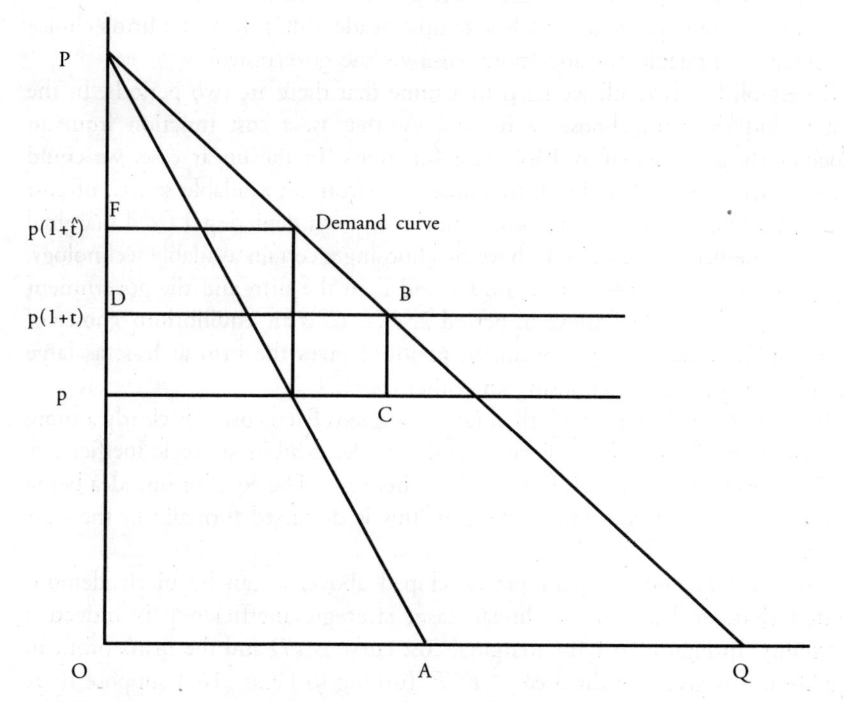

Figure 16.3

will intersect the marginal-cost curve at point G. Thus, the firm's optimal choice of producer price is p^*.

Next, we examine the government's choice of t given producer price p. In Figure 16.3, the producer price is p. If the government sets the tax rate at t, its revenue collection will clearly be equal to $PDBC$. For a given producer price p, the tax revenue is given by a rectangle sitting on the line pC and with a corner on the demand curve PQ. Clearly, this is maximized where the marginal-revenue curve, PA, intersects pC, at the tax rate $\hat{t}$ where $\hat{t}$ is implicitly defined by $(1+\hat{t})p = OF$. It is immediate that if the producer price is set at p^* in Figure 16.1, the government's optimal choice of tax rate t^* is equal to $(p_m^*/p^*) - 1 = (OP/OB) - 1$. This completes the proof.

It is easy to see that $BE = ED$ is also a *necessary* condition for equilibrium. It is obvious from this and from an inspection of Figure 16.1 that the equilibrium is unique in the linear case.

Using this diagrammatic characterization some properties of the linear case are immediate.

(1) If a firm has a lower marginal cost of production, it will face a higher ad valorem tax rate, charge a lower price, and produce a larger amount.

(2) Equilibrium output is less than what would have been produced by a traditional monopolist and the consumer's price higher.

We can now proceed to give a first view of the strategic-inefficiency result. It will be shown that when a firm plays strategic games with the government, as in the above model, it may be in the firm's interest to be inefficient.

Unlike in some existing models (example Seade 1987), here the firm's choice is guided by strategic considerations vis-à-vis the government.

To establish this result we need to assume that there are two periods. In the first period the firm chooses a technology, that is, a cost function from an exogenously given set of available cost functions. In the linear case, we could equivalently assume that the firm chooses one from an available set, $\bar{C}$, of cost functions. Thus, $\bar{C}$ is a set of positive numbers, each depicting a fixed marginal cost that the firm can choose to have by choosing a certain available technology.

The firm chooses $c \in \bar{C}$ in period 1 and then the firm and the government play the game described above in period 2. $c^* \in \bar{C}$ is an 'equilibrium' choice of firm 1 if the resulting equilibrium in period 2 gives the firm at least as large a profit as it can get by choosing any other $c \in \bar{C}$.[3]

If $c, c' \in \bar{C}$ such that $c > c'$, then (assuming zero fixed cost) c is clearly a more inefficient technology. We will say that the model exhibits strategic inefficiency if $\bar{C} = \{c, c'\}$ such that $c > c'$, but the firm chooses c. The equilibrium idea being used here is that of subgame perfection; this is discussed formally in the next section.

Using the geometric technique developed above, it can be nicely demonstrated that, at least in the linear case, strategic inefficiency is indeed a possibility. In Figure 16.1 the marginal-cost curve is FD and the firm's profit in equilibrium is given by the area $p^* EGF$. Turning to Figure 16.4 suppose there is a new marginal-cost curve, $F'D'$, which is lower than FD. Since $BE = ED$, it

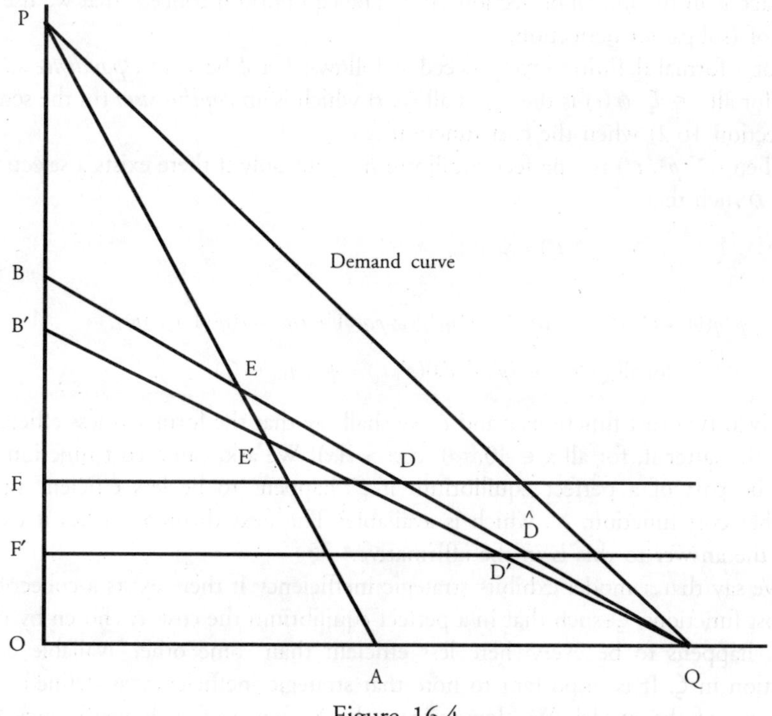

Figure 16.4

follows that $BE < E\hat{D}$. Hence, by the bisection rule, the new equilibrium will be on a flatter ray, QB', such that $B'E' = E'D'$.

Now consider a sequence of increasingly efficient technologies that lower the marginal-cost curve. In particular, consider a sequence of constant marginal-cost curves converging to the zero marginal-cost curve.

Since $OA = AQ$, as marginal cost goes towards zero, the equilibrium ray clearly goes towards OQ. Further note that for marginal cost FD, the firm's profit is less than the area OBQ. But as marginal cost goes to zero, BQ tends to OQ and hence the firm's profit goes towards zero.

Therefore, we can pick a marginal-cost curve from the sequence of curves going towards zero, which generates a smaller profit than $p^* EGF$ in Figure 16.1. Hence, if the only two technologies available to the firm are the one generating the marginal-cost curve we just picked, and the one that generated FD in Figure 16.1, the firm will pick the latter even though it is more inefficient.

16.4. Strategic Inefficiency: A General Theorem

The aim of this section is to generalize the strategic-inefficiency claim of the previous section. As before, in period 1, the firm chooses a technology, embodied here entirely in a cost function, from an available set ζ, of cost functions. In period 2, with the cost function given, the firm and government interact as in the model of Section 16.2. The equilibrium concept that we use is that of (subgame) perfection.

For a formal definition we proceed as follows: Let ϕ be a *correspondence* such that for all $c \in \zeta$, $\phi(c)$ is the set of all (p, t) which is an *equilibrium* (in the sense of Section 16.2) when the cost function is c.

Then (c^*, p^*, t^*) is a perfect equilibrium if and only if there exists a selection ψ in ϕ such that

$$(p^*, t^*) = \psi(c^*),$$

and

$$p^*q((1 + t^*)p^*) - c^*(q((1 + t^*)p^*)) \geq pq((1 + t)p) - c(q((1 + t)p)),$$
$$\text{for all } (p, t, c) \in (p', t', c')|(p', t') = \psi(c'), c' \in \zeta).$$

Given two cost functions c and $\hat{c}$, we shall say that the former is less efficient than the latter if, for all $x \in [0, \infty)$, $c(x) > \hat{c}(x)$. We ask, can a cost function c^* ever be part of a perfect equilibrium if c^* happens to be less efficient than another cost function, c^*, which is available? The next theorem makes it clear that the answer to this is in the affirmative.[4]

We say that a model exhibits strategic inefficiency if there exists a collection of cost functions, ζ, such that in a perfect equilibrium the cost, c, chosen by the firm, happens to be everywhere less efficient than some other available cost function in ζ. It is important to note that strategic inefficiency as defined is a property of the model. We demonstrate the existence of such inefficiency by restricting attention to the class of linear cost functions. Hence the set, ζ, of admissible cost functions is a subset of $\{c(\cdot)|\ \forall x, c(x) = c \cdot x, \text{ where } c \in (0, \infty)\}$. We shall refer to these cost functions by their (constant) marginal cost.

THEOREM 16.2: Models of government–firm interaction satisfying assumptions (i)–(v) exhibit strategic inefficiency.

PROOF: This is broken up into three steps.

Step 1. Let us first define surrogate profit and revenue functions

$$\tilde{\pi}(q, t, c) = \left[\frac{r(q)}{1 + t} - c\right] q, \tag{16.4}$$

and

$$\hat{R}(p, q) = R\left(p, \frac{r(q)}{p} - 1\right) = \left[\frac{r(q)}{p} - 1\right] pq. \tag{16.5}$$

It is easy to check that $(\hat{p}, \hat{t})$ is a Nash equilibrium associated with $\hat{c}$, if and only if there exists a $\hat{q}$ such that

$$r(\hat{q})/(1 + \hat{t}) = \hat{p},$$

and

$$\hat{q} \in \underset{q \in [0, Q]}{\operatorname{argmax}}\ \hat{R}(\hat{p}, q) \cap \underset{q \in [0, Q]}{\operatorname{argmax}}\ \tilde{\pi}(q, \hat{t}, \hat{c}). \tag{16.6}$$

This alternative characterization of the Nash equilibrium in the second period is now used to complete the proof.

Step 2. We now prove that for all $p \in (0, P)$, there exist c and t such that $0 < c < p$, and $(p, t) \in \phi(c)$. Let $0 < \hat{p} < P$ and choose any

$$\hat{q} \in \underset{q}{\operatorname{argmax}}\ \tilde{R}(\hat{p}, q). \tag{16.7}$$

Define

$$\hat{t} \equiv (r(\hat{q})/\hat{p}) - 1,$$

and

$$\hat{c} \equiv \hat{p}^2 / r(\hat{q}).$$

Then it will be shown that $(\hat{p}, \hat{t}) \in \phi(\hat{c})$. In the light of (16.7) and the alternative Nash-equilibrium characterization in step 1, all we need to show is that

$$\hat{q} \in \operatorname{argmax}\ \tilde{\pi}(q, \hat{t}, \hat{c}). \tag{16.8}$$

Note that (16.8) is equivalent to

$$\left[\frac{r(\hat{q})}{1+t} - \hat{c}\right]\hat{q} \geq \left[\frac{r(q)}{1+t} - \hat{c}\right]q, \text{ for all } q,$$

or

$$r(\hat{q})\hat{q} - r(q)q + \hat{p}(q - \hat{q}) \geq 0. \tag{16.9}$$

If we go through the same process of writing out (16.7) in long-hand we arrive at (16.9). Hence (16.7) implies (16.8).

Step 3. Step 2 implies we can construct a sequence $\{(c_n, p_n, t_n)\}$ such that, for all n, $c_n > 0$, $(p_n, t_n) \in \phi(c_n)$ and $p_n > 0$; and $\lim c_n = 0$ and $\lim p_n = 0$. Clearly then, the firm's profit goes to zero as c_n goes to zero. It is easy to check that for some $c > 0$, the firm earns a positive profit in Nash equilibrium. Hence, we could construct a ζ such that the firm chooses an inefficient cost function from ζ in the perfect equilibrium. Q.E.D.

Strategic inefficiency may not occur if the demand function does not satisfy the assumptions in Section 16.2. This is true, for instance, with all constant-elasticity demand curves.[5] Since in such a case $q(0)$ is not defined, such cases lie outside our framework. Essentially, what seems to be the (rather weak) requirement for strategic inefficiency to occur is to have demand curves such that (a) the marginal revenue (if defined) 'is zero at some finite output', and (b) there exists a high enough P for which $q(P)$ happens to be zero.

16.5. Stackelberg Equilibria

In this section, we examine alternatives to the Nash characterization of the second-stage game. The two obvious cases are the two Stackelberg games: S^1, where the government first chooses t, followed by the firm's choice of p; and S^2,

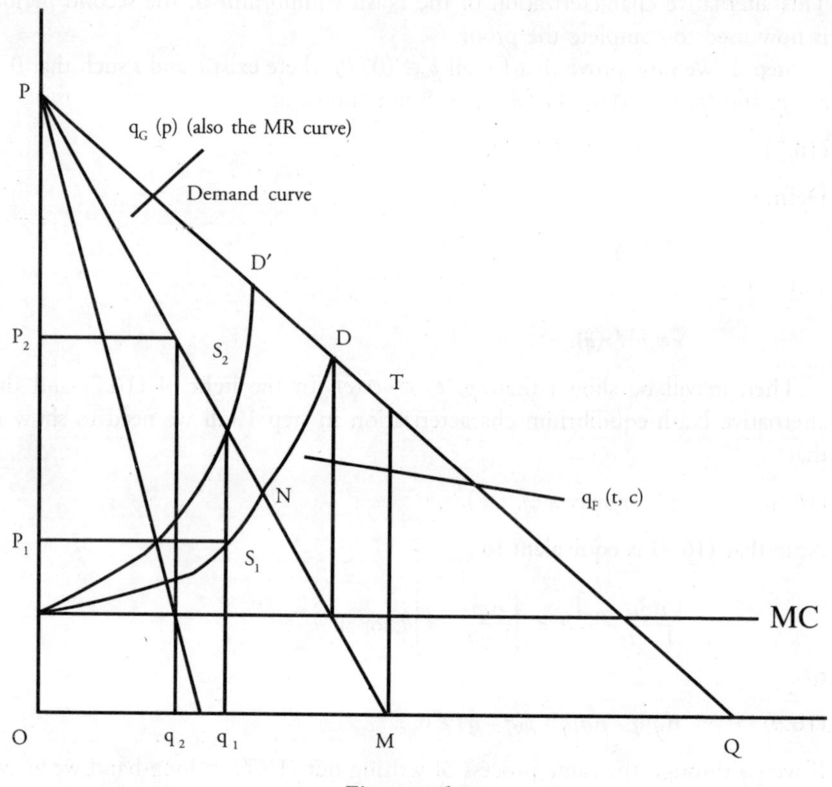

Figure 16.5

the opposite sequence. First, consider the linear example of Section 16.3. Using the arguments of Section 16.3 we can describe the firm's behaviour by a surrogate reaction function $q_F(t, c)$ which specifies the firm's choice of quantity for every (t, c). The firm actually chooses price but, given t, quantity and price have a one-to-one relation. Similarly $q_G(p)$ is the government's surrogate reaction function.

From our earlier discussion, we know that $q_G(p)$ coincides with the marginal revenue curve. The $q_F(t, c)$ curve can be derived geometrically. With c constant, for every post-tax demand curve, we determine the profit-maximizing price–quantity pair in the usual manner. The locus of these points for different values of t is illustrated in Figure 16.5 and slightly incorrectly labelled as the $q_F(t, c)$ curve. It is increasing and convex. It begins where the marginal cost touches the vertical axis and ends at the monopoly equilibrium on the demand curve.[6]

Consider first S^2. For each choice of p by the firm, the government will maximize revenue by choosing the corresponding point on the marginal-revenue curve PM. The maximum profit for the firm can be located at S_2 by treating PM as the demand curve in a conventional monopoly. Price will be p_2 and tax will be given by the post-tax demand curve that goes through S_2.

S^1 is more complicated. The government knows that, given the firm's reaction function, it will invariably end up on curve *CD*. For each point on *CD* the government's revenue is equal to the area given by the rectangle formed by the demand curve and the vertical axis. So the government's problem is analogous to that of an ordinary monopoly facing the demand curve *PQ* and average-cost curve, *CD*. The solution uses the marginal curve (shown as *CD'*) to this pseudo 'average-cost' curve, *CD*. The intersection of CD' with marginal revenue *PM* identifies the equilibrium. The equilibrium tax rate is given by the t-demand curve going through point S_1 with price as P_1. The Nash equilibrium is the (point N) intersection of curves *CD* and *PM* in Figure 16.5. It follows that with the government as Stackelberg leader the equilibrium tax rate will be greater than in the Nash equilibrium; price will be less and quantity produced would be less as well.

Supposing, now, that before playing each of these Stackelberg games the firm can commit itself to a cost function. It is easy to check that the inefficiency result does not hold in case S^2. But as we turn our attention to S^1, the strategic-inefficiency result comes back, as seen in Theorem 16.3. This is interesting given that S^1 is close to the standard characterization of government–firm interactions.

THEOREM 16.3: The perfect equilibria of three-period models of government–firm interaction, satisfying assumptions (i)–(v), in which choices are made in the sequence, (c, t, p), exhibit strategic inefficiency.

PROOF. Let $R^N(c)$ be the maximum revenue that the government can earn in a Nash equilibrium when the (constant) marginal cost of production is c. That is. $R^N(c)$ solves

$$\text{Max } R(p, t)$$

subject to

$$(p, t) \in \phi\ (c)\ .$$

It follows that if p is the price that prevails in a Nash equilibrium under c, then

$$R^N(c) \geq \max_q \tilde{R}(p, q) = \max_q\ [r(q)q - pq]. \tag{16.10}$$

Next, note that if $\{p_n\}$ is a sequence converging to zero, then

$$\max_q\ [r(q)q - p_n q] = \max_q r(q)q. \tag{16.11}$$

We know from the proof of Theorem 16.2 that there exists a sequence $\{(c_n, p_n, t_n)\}$ such that $(p_n, t_n) \in \phi\ (c_n)$, $c_n \to 0$ and $p_n \to 0$. It follows from (16.10) and (16.11) that

$$\lim_{n \to x} R^N(C_n) \geq \max_q r\ (q)q\ .$$

But R^N can never exceed max $r\ (q)q$ since that is the total profit in the system. Hence, lim $R^N(c_n) = \max_q r\ (q)q$. Now, given marginal cost c, let $R^S(c)$ be the revenue earned by the government if the government plays Stackelberg leader. Clearly, for all c, $R^S(c) \geq R^N(c)$. Hence $R^S(c) = \max_q r(q)q$.

Let the firm's profit in this same Stackelberg be $\pi^S(c)$. Since the total of profits and taxes cannot exceed max $r(q)q$ we have $\lim \pi^S(c_n) = 0$. Since there exist c: $\pi^S(c) > 0$, it follows that there will exist two cost functions c' and c'' such that the firm chooses the less efficient one. Q.E.D.

16.6. Extensions

In this section we examine the robustness of the results to relaxing some of the key assumptions of the models. Namely: (a) Would the inefficiency results be valid if the government's objective is not merely to maximize revenue but something more complicated like a weighted average of revenue and aggregate welfare? (b) How would the results be affected if instead of an ad valorem tax the government had to choose a specific or unit tax? and (c) Is there any reason to believe that in the context described in the chapter the government would in fact use an ad valorem tax?

It will be seen that the results are indeed generalizable to such modifications. For (a) we have investigated several alternative formulations and the answer is 'yes'. We do not present the algebra here (it being available on request).

Turning to (b) and (c), unit taxes are easy to analyse and it can clearly be seen that inefficiency cannot arise in this case. However, the answer to (b) turns out to be less consequential in the light of the answer we get to (c). If we model government's choice of the type of tax, ad valorem versus per unit, as part of the 'game' then we find that in equilibrium the government will choose to use an ad valorem tax. So, though the unit tax averts the strategic-inefficiency problem, in the government–firm game modelled here the government will use an ad valorem tax.

To begin with (c), we use the assumptions of Section 16.2 and, in addition, assume that the demand function is continuously differentiable (which considerably shortens the proofs). Consider a model where the government chooses between using A, an ad valorem tax and U, a unit tax. We denote this choice by a variable, d, that takes values A or U, then it chooses the tax rate, t_d, and this is then followed by the firm's choice of price, p.

Given a choice, (d, t_d, p) the firm's profit is given by

$$\pi(A, t_A, p) \equiv \pi(p, t_A) \equiv pq((1 + t_A)p) - c(q((1 + t_A)P)),$$

and

$$\pi(U, t_U, p) \equiv \bar{\pi}(p, t_U) \equiv pq(p + t_U) - c(q(p + t_U)),$$

depending on whether $d = A$ or U.

The government's revenue is given by

$$R(A, t_A, p) \equiv R(p, t_A) \equiv t_A pq((1 + t_A)P),$$

and

$$R(U, t_U, p) \equiv \bar{R}(p, t_U) \equiv t_u q(p + t_U).$$

Note that a bar denotes the case of a specific or unit tax.

THEOREM 16.4: In perfect equilibrium the government uses an ad valorem tax, i.e., $d = A$.

PROOF: Define

$$P(t_A) = \underset{p}{argmax}\ \pi(p, t_A), \tag{16.12}$$

$$\bar{P}(t_U) = \underset{p}{argmax}\ \bar{\pi}(p, t_U). \tag{16.13}$$

Step 1. If $t_A{}^*$ and $t_U{}^*$ are such that

$$p(t_A{}^*)(1 + t_A{}^*) = \bar{p}(t_U{}^*) + t_U{}^*, \tag{16.14}$$

then $\bar{p}(t_U^*) > p(t_A^*)$. From (16.12) and (16.13) we know that

$$\frac{\partial \pi}{\partial p}(p(t_A{}^*), t_A{}^*) = 0 \text{ and } \frac{\partial \bar{\pi}}{\partial p}(\bar{p}(t_U{}^*), t_U{}^*) = 0.$$

Writing these out in full we get (omitting the asterisks):

$$q((1 + t_A)p(t_A)) + p(t_A)q'((1 + t_A)p(t_A))(1 + t_A) - c'(\cdot)q'(\cdot)(1 + t_A) = 0,$$

and

$$q(\bar{p}(t_U) + t_U) + \bar{p}(t_U)q'(\bar{p}(t_U) - t_U) + c'(\cdot)q'(\cdot) = 0.$$

This, in turn, implies that

$$p(t_A) = \frac{c'(\cdot)q'(\cdot)(1 + t_A) - q(\cdot)}{q'(\cdot)(1 + t_A)}, \tag{16.15}$$

$$\bar{p}(t_U) = \frac{c'(\cdot)q'(\cdot) - q(\cdot)}{q'(\cdot)}. \tag{16.16}$$

Then (16.14) implies that

$$q((1 + t_A)p(t_A)) = q(p(t_U) + t_U),$$

and

$$q'((1 + t_A)p(t_A)) = q'(p(t_U) + t_U),$$

For all t_A and t_U satisfying (16.14), $q'(\cdot)$ and $q(\cdot)$ in (16.15) and (16.16) are the same: therefore, $\bar{p}(t_U) > p(t_A)$. This establishes the claim in step 1.

Step 2. Next, if t_A and t_U satisfy (16.14), then

$$\bar{R}(\bar{p}(t_U), t_U) < R(p(t_A), t_A).$$

This follows as an immediate consequence of $\bar{p}(t_U) > p(t_A)$, which was proved in step 1.

Step 3. It will now be proved that

$$\max_{t_A} R(p(t_A), t_A) > \max_{t_U} \bar{R}(\bar{p}(t_U), t_U). \tag{16.17}$$

Let

$$\hat{t}_A \equiv \underset{t_A}{\text{argmax}}\ R(p(t_A), t_A),$$

and

$$\bar{t}_U = \underset{t_U}{\text{argmax}}\ \bar{R}(\bar{p}(t_U), t_U).$$

Let t_A be such that

$$p(t_A^*)(1 - t_A^*) = \bar{p}(\bar{t}_U)\ \bar{t}_U.$$

Hence,

$$\bar{R}(\bar{p}(\bar{t}_U), (\bar{t}_U) < R(p(t_A^*), t^*), \text{ by step 2}$$
$$\leq R(p(\hat{t}_A), \hat{t}_A), \text{ by the definition of } \hat{t}_A.$$

This proves step 3. It is immediately obvious that choosing an ad valorem tax and setting the rate equal to $\hat{t}$ (and earning $R(p(\hat{t}), \hat{t})$ is part of a perfect equilibrium. Q.E.D.

In a game-theoretic model of industry, the sequence of moves often turns out to be critical for the results. What is surprising is that the above result is robust to some reasonable alterations of the sequence. The above theorem pertains to the case where, first (1) costs are chosen; next (2) the form of tax (i.e., A or U) is chosen by the government; then (3) the government chooses the tax rate (t_A or t_U); and, finally, (4) the firm chooses the price. What happens, it may be asked, if the firm chooses cost between (2) and (3)? In other words, the sequence of moves is (2), (1), (3), and (4). It is easy to see that for a collection of cost functions C the theorem still goes through; that is, the ad valorem tax still emerges in equilibrium: Theorem 16.4 established this result for any cost function, say, c. Now take a collection of cost functions in the neighbourhood of c. Call the collection C. For a small enough neighbourhood C, even with sequence (2), (1), (3), and (4), the ad valorem system will be chosen in equilibrium because in the limiting case, where $C = \{c\}$, this sequence of moves is indistinguishable from the sequence (1), (2), (3), and (4), and the strict inequalities in the proof will continue to hold; and further, in the latter case, we know from Theorem 16.4 that the ad valorem tax prevails.

APPENDIX 16.1

The purpose of this appendix is to prove Theorem 16.1. The proof is given by using a well-known variant of Nash's existence theorem for one-shot non-cooperative game.

In what follows, we use E to denote the set of real numbers and E_+ its non-negative subset. Lemma A16.1 states a property of composite functions. It is a special case of a more general mathematical proposition, trimmed suitably for our purpose.

LEMMA A16.1: Suppose a mapping $f: E_+ \to E_+$ had the property that for all $x, y \in E_+$, if $x \geq y$, then $f(x) \leq f(y)$. Let $f(E_+)$ be a subset of the interval $[a, b]$. If the mapping $g: [a, b] \to E$ is concave, then the composite mapping $g \cdot f$ must be quasi-concave.

PROOF: Suppose the hypothesis of the lemma is true.

Let x and y be real numbers and $x < y$ and let $h \in [0,1]$. Since f is negative monotonic

$$f(x) \geq f(hx + (1 - h)y) \geq f(y).$$

Hence, there exists $\alpha \in [0,1]$ such that

$$f(hx + (1-h)y) = \alpha f(x) + (1-\alpha) f(y). \tag{A16.1}$$

Since g is concave, g must be quasi-concave. Hence

$$g(\alpha f(x) + (1-\alpha) f(y)) > \min\{g(f(x)), g(f(y))\}. \tag{A16.2}$$

(A16.1) and (A16.2) imply

$$g \cdot f(hx + (1-h)y) \geq \min\{g \cdot f(x), g \cdot f(y)\}.$$

LEMMA A16.2: The mapping $\pi(p, t)$ is quasi-concave in p and the mapping $R(p, t)$ is quasi-concave in t.

PROOF: Define a new function $\tilde{\pi}$ as follows

$$\tilde{\pi}(x, t) \equiv \pi\left(\frac{r(x)}{1+t}, t\right) = \frac{xr(x)}{1+t} - c(x).$$

Clearly, $\tilde{\pi}$ is concave in x, since $xr(x)$ and $-c(x)$ are both concave.

Given the definition of $\tilde{\pi}$, it follows that

$$\pi(p, t) = \tilde{\pi}(q((1+t)p), t).$$

Hence, for a given t, π is a composite mapping of $\tilde{\pi}$ and q. Since q is inversely monotonic and $\tilde{\pi}$ is concave in the first variable, by Lemma 1 we know that $\pi(p, t)$ is quasi-concave in p.
The quasi-concavity of $R(p, t)$ in t is proved in the same manner by defining the following new function:

$$\tilde{R}(p, x) = R\left(p, \frac{r(x)}{p} - 1\right) = \left(\frac{r(x)}{p} - 1\right) pq. \qquad \text{Q.E.D.}$$

PROOF OF THEOREM 16.1. Let b be a positive real number such that for all $x \in [0, \infty)$, $c(x) \geq bx$. The existence of such a b is ensured by assumption (v) in Section 16.2. Define

$$T = \frac{P}{b} - 1.$$

[Recall the definition of P from (i) in Section 16.2.]

Suppose now the firm and the government are playing a game of maximizing respectively π and R. The firm chooses a p from the interval $[b, P]$ and the government chooses a t from the interval $[0, T]$.

Note that now

(A) each player chooses his strategy from a convex and compact set,
(B) the mappings π and R are continuous since q and c are continuous, and
(C) π is quasi-concave in p and R is quasi-concave in t, by Lemma 2.

By a well-known variant of Nash's theorem (see Friedman 1977: 160, Theorem 7.4), it follows that this game must have a Nash equilibrium. That is, there exists $(p^*, t^*) \in [b, P] \times [0, T]$ such that

$$\pi(p^*, t^*) \geq \pi(p, t^*), \quad \text{for all } p \in [b, P]. \tag{A16.3}$$

and

$$R(p^*, t^*) \geq R(p^*, t), \quad \text{for all } t \in [0, T]. \tag{A16.4}$$

The only thing that remains to be proved is that the domain restrictions (price to $[b, P]$, and tax rate to $[0, T]$) are inessential: suppose (p^*, t^*) satisfy (A16.3) and (A16.4). Consider a $p \in [0, \infty)\backslash[b, P]$. If $p > P$, then $q((1 + t^*)p) = 0$. Hence, $\pi(p, t^*) = 0$. Suppose, now, $p < b$. Note that

$$\pi(p, t^*) < pq((1 + t^*)p) - bq((1 + t^*)p) < 0$$

by the definition of b, since $p < b$. Since $\pi(p, t^*) = 0$, we know by (A 16.3) that

$$\pi(p^*, t) \geq 0.$$

Hence,

$$\pi(p^*, t^*) \geq \pi(p, t^*), \qquad \text{for all } p \in [0, \infty)\backslash[b, P].$$

Therefore, together with (A16.3), we have

$$\pi(p^*, t^*) > \pi(p, t^*), \qquad \text{for all } p \in [0, \infty).$$

Suppose $t > T$. Then

$$(1 + t)p^* > (1 + T)b, \text{ since } p^* \geq b,$$

$$= P, \text{ by the definition of } T.$$

This implies $q((1 + t)p^*) = 0$.

Hence, $R(p^*, t) = 0$, by (16.2). Since $R(p, t)$ is non-negative for all $(p, t) \in [0, \infty) \times [0, \infty)$, it follows, $R(p^*, t^*) \geq R(p, t)$, for all $t > T$.

This and (A16.4) imply

$$R(p^*, t^*) \geq R(p^*, t), \qquad \text{for all } t \in [0, \infty). \text{ Q.E.D.}$$

NOTES

1. This inefficiency is different from the technical and allocative inefficiencies studied in the literature e.g., Farrell (1957). The cost-raising feature is common to both types of inefficiency. Strategic inefficiency as discussed in this section is distinct from slack or X-inefficiency (see, e.g., Selten 1986; Leibenstein 1987).
2. In the light of the convexity assumption this implies that marginal cost is bounded away from zero.
3. Similar ideas have been used in models of entry-deterrence (Spence 1977; Dixit 1980; Basu and Singh 1990).
4. Such inefficiency can be explained in two ways: either as the direct outcome of profit maximization, or we could think of different firms as following different customary rules of behaviour and we use a natural selection argument, which is becoming popular in the literature (see Jacquemin 1987), to explain why only the inefficient ones would survive in an environment described by our model.
5. While this may be worked out algebraically, there is a simpler explanation due to the fact that with ad valorem taxes the post-tax demand continues to be of the same elasticity. Hence, since the monopoly price–cost margin is determined by the elasticity of demand, the price is independent of the tax rate.
6. In the general, non-linear case the shape of this could be quite irregular, violating monotonicity and continuity. We could have relaxed the assumption of linearity to concavity and $q_F(t, t)$ would be quite similar.

References

Basu, K. and N. Singh. 1990. 'Entry-deterrence in Stackelberg Perfect Equilibria'. *International Economic Review* 30: 61–71.

Casson, M. 1987. *The Firm and the Market*. Oxford: Basil Blackwell.

Dixit, A. 1980. 'The Role of Investment in Entry Deterrence'. *Economic Journal* 90: 95–106.

Farrell, M.J. 1957. 'The Measurement of Productive Efficiency'. *Journal of the Royal Statistical Society*. Series A. General. 120(3): 253–81.

Friedman, J. 1977. *Oligopoly and the Theory of Games*. Amsterdam: North-Holland.

Jacquemin, A. 1987. *The New Industrial Organization*. Cambridge, M.A.: MIT Press.

Kindleberger, C.P. 1984. *Multinational Excursions*. Cambridge, M.A.: MIT Press.

Leibenstein, H. 1987. *Inside the Firm. The Inefficiencies of Hierarchy*. Cambridge, M.A.: Harvard University Press.

McGuire, T., M. Coiner, and L. Spancake. 1979. 'Budget Maximizing Agencies and Efficiency in Government'. *Public Choice* 34: 333–57.

Niskanen, W. 1971. *Bureaucracy and Representative Government*. Chicago, IL: Aldine.

Seade, J. 1987. 'Profitable Cost Increases and the Shifting of Taxation: Equilibrium Responses of Markets in Oligopoly'. Mimeo.

Selten, R. 1986. 'Elementary Theory of Slack-ridden Imperfect Competition'. In G.F. Mathewson and J. Stiglitz, eds. *New Developments in the Analysis of Market Structure*. Cambridge, M.A: MIT Press.

Spence, A.M. 1977. 'Entry, Capacity, Investment and Oligopolistic Pricing'. *Bell Journal of Economics* 8: 534–44.

Stern, N. 1987. 'The Effects of Taxation, Price Control and Government Contracts in Oligopoly and Monopolistic Competition'. *Journal of Public Economics* 32: 133–58.

von Furstenberg, G.M., R.J. Green, and J. Jeong. 1986. 'Tax and Spend or Spend and Tax'. *Review of Economics and Statistics* 68: 179–88.

17 On Misunderstanding Government

An Analysis of the Art of Policy Advice

17.1. On Misunderstanding Government

There are situations in life where individuals, left to themselves, create inefficiency and anarchy. This was the concern that drove Hobbes to philosophize about the Leviathan that can bring order into chaos; and this is a concern that has provoked much contemporary writing. The prisoner's dilemma is the classic description of this predicament (see Taylor 1976). It codifies how in some situations individuals, left to pursue their individual interests, can harm themselves. The prisoner's dilemma is an illustration of extreme externality between individuals. Smaller externalities are pervasive in life. They form the basis of our contemporary environmental concerns. If individuals work in their atomistic interest, then each individual may ignore the pollutants their actions inject into the atmosphere or the little damage that their factories may do to the ozone layer. However, the totality of these actions can leave all individuals worse off with severe environmental damage.

What should be done about this problem? To many the answer is obvious. The above discussion shows how individual rationality may fail to be an adequate organizing force in society and so what we need is government intervention to bring *individual* actions into alignment with *social* interests. Thus what the two prisoners in the prisoner's dilemma need is a third party to help them maximize their own welfare (by, for instance, punishing the person who defects and confesses).

This advice for government intervention is flawed for a non-obvious reason.

From *Economic and Politics* 9 (November), 1997: 231–50.

I have benefited from conversations with Joe Halpern, Peter Katzenstein, Andrew Rutten, Eduardo Saavedra, Nirvikar Singh, Jorgen Weibull, and Eduardo Zambrano.

If the advice were followed it would bring about the kind of world being recommended. So in *that* sense there is nothing wrong with the advice. What is wrong is that it is unlikely to be followed, because it takes an excessively simplistic view of government. Government is treated as a benevolent agent, exogenous to the economy, whose intervention can be invoked at will. It ignores the fact that government is itself a collection of individuals with their own motivations and aspirations.

Likewise, as we just saw, dismayed by the outcome of the prisoner's dilemma, economists and political scientists often recommend that what is needed is a 'third party' to induce individuals to behave in their group interest. The problem with this remedy is that if there is such a third party around then we had no right to model the game as a two-person prisoner's dilemma in the first place. The agent who is willed out of thin air should have been a part of the game to start with. If the presence of the third person means that we do not have a prisoner's dilemma then we never had a prisoner's dilemma; and if even with the third person included we still have a prisoner's dilemma then that *is* our plight—there is no escape from it.

In this same spirit the persons who comprise government should, strictly, be thought of as players in the 'economy game'. In his recent monograph on the political process, Dixit (1996:2) takes a similar line: 'Most important, I will argue that the political process should be viewed as a game ... What follows from these observations is orthogonal to, and perhaps destructive of, the whole "markets versus governments" debate'.

If prices are high very few economists today would say that their advice is that the producers should lower the price. They do not give this advice not because the advice is wrong but because they feel it is a futile advice since producers have their own objectives and will not heed such advice. Hence, the profusion of advice that economists target at government reflects, more than anything else, how little they understand government.

Our poor understanding of government has showed up in other places too. Some economists argue that individuals should be entirely free to pursue their own interest, free from government intervention. This misses the point that government is itself an assemblage of individuals.

Similarly some economists believe that institutions that have emerged out of individual actions are optimal; they are there because their benefits outweigh the costs (Anderson and Hill 1975; Posner 1981). Government, according to this view, is an organization that distorts these natural civic institutions. Such a view is made possible only by not asking ourselves where government itself has come from. Government did not create government; so we cannot castigate it as one more manifestation of the evils of government. Indeed government is a fairly modern institution (Strayer 1970). It has evolved through the ages, through a multitude of individual actions. (We return to some of these issues in Section 17.6.) Hence, if we maintain that individuals, left to themselves, bring about desirable institutions, then we cannot say that such institutions need to be protected from government because government is such an institution.

Moreover, to take the view that it is all right for individuals to bring about institutions which help them co-operate and enhance their well-being but that they should not create the Leviathan amounts to placing an exogenous constraint on *individual* effort. It is the fallacy that Newt Gingrich commits when he says that government should get out of charitable and welfare work and leave these to community efforts. What this misses out on is that government itself is one such community effort.

To see the intricacy of this argument and the ease with which we can err in our conception of government, let us consider Taylor's celebrated critique of Hobbes' (1651) justification for strong government. As Taylor observes in his *Preface,* in the West the most popular justification of the state is that 'without the state, people would not act so as to realize their common interests; more specifically, they would not voluntarily cooperate to provide themselves with certain public goods'. The roots of this view, he argues, go back principally to Hobbes (and also to Hume). Taylor's book is meant to be a critique of this view. He is right in challenging this view, which is based on the 'third-party' (or exogenous or 'puppet') view of government which we have already taken to task. However, one does not have to be a careful reader to see that his criticism is very different from the one made earlier in this section. Taylor shows how individuals can voluntarily co-operate, for instance, through repeated play of the prisoner's dilemma. This then becomes the basis of his rejection of the Hobbesian recommendation whereby individuals create the Sovereign who then ensures order and co-operation among the citizens by 'creating appropriate laws and punishing transgressors' (Taylor 1976: 104).

The error in this viewpoint is to characterize the government-led path to order as a 'coercive one' and the repeated prisoner's dilemma path as the 'voluntary one'. Hence, Taylor's critique also commits' the fallacy of the exogenous conception of government. Since, in reality, government is itself a creation of the individuals and is run by the individuals, the government-led path to social order is nothing but a self-enforcing equilibrium among the individuals—a point that is made with great ingenuity by David Friedman (1994). After all, even in the repeated prisoner's dilemma' co-operation arises from the threat of punishment to deviators from the co-operative path. Just because in the Hobbesian route the 'punishers' are *called* members of government this does not make the Hobbesian path any more coercive than the one sustained by a trigger strategy in a repeated game.

In brief, we cannot maintain that (a) individuals are always rational, (b) institutions that emerge from the actions and choices of individuals are desirable, and (c) big governments are undesirable.

Much of the popular debate concerning 'big' and 'small' governments is of such intellectually poor quality precisely because it is rooted in this fallacy. We can argue that governments are often too big and too oppressive but we have to construct such an argument on the negation of (a) or (b).

The endogenization of government is a large research agenda. The works of Buchanan (1968), Taylor (1976), Bhagwati et al. (1984), Dixit (1996), and other contemporary writers have contributed building blocks to this project. In the next four sections, I also focus on a small part of this big agenda. The question that is addressed is this: Once we take an endogenous, game-theoretic view of government, how do we, as economists or political scientists, advise such a government? This question belongs to the genre of problems raised in O'Flaherty and Bhagwati (1996). I will argue that the conventional view on *how* we should give advice is simplistic and inadequate, being predicated on a flawed conception of government. Once this is corrected, advising governments turns out to be a subject of considerable intellectual challenge, which has the potential of bridging the gap between research and actual policy.

17.2. A Science of Advising?

Viewed through Machiavellian lenses, the economist's faulty conception of government is not so much an act of folly as a ruse or an act of strategem. It is the existence of such a 'puppet government' (Bhagwati 1990; Srinivasan 1992) or what Milton Friedman (1986) called a 'public interest' conception of government that justifies the existence of the traditional policy adviser. And economists like to give advice. As O'Flaherty and Bhagwati (1996) observe, 'Many economists like to think of themselves as active participants in history, not as members of a contemplative order trying to understand a world they cannot influence'.

But a strategy constructed over such severe fault lines, as discussed in the previous section, cannot work too well. And indeed economists, and, more generally, social scientists have faltered when it has come to the 'how' of giving advice.

It is a pity that there does not exist a science of advising. The accumulated expertise and knowledge of the social sciences have failed to translate adequately into action because we have neglected the study of the *process* of transmission of information from the expert (the adviser) to the minister, senator, or politician (more generically, the advisee) who is responsible for putting plans into action.

Our traditional conception of government is not only empirically flawed, as argued above, but it also has problems of internal consistency with the standard model of economics.

Consider the Arrow-Debreu model of general equilibrium. In it individuals choose points from their budget sets, buy and sell goods; and out of this emerges what are called equilibrium prices. In this economy what person *i says* does not affect what person *j does*. In brief, it is an economy that works in silence. This is not to deny that in the Arrow-Debreu world people may be chatting, laughing, and singing, but it is simply that that aspect of their lives does not impinge on what happens in the domain of economics and in the market place.

Having studied such a model in which *i*'s speech has no effect on what *j* does, the policy economist has gone on to give advice. But what is advice but a set of spoken or written words? If the advice is based on a model in which such words cannot have any affect, then it is surely inconsistent to believe that the advice can have an effect.

This, in essence, is closely related to the 'determinacy paradox' of Bhagwati et al. (1984). Once we endogenize all agents, the system we are studying may become fully determined leaving no scope for the policy adviser. While the mainstream of economics has overlooked this paradox, there is a small body of writing that has tried to contend with this deep problem.[1]

There are two broad ways in which a piece of advice can be considered defective. First, advice can be 'wrong advice' in the sense of it being based on an erroneous view of the world so that *if* it were followed, it would not bring about the kind of world it was intended to bring about. Second, advice can be 'futile advice' in the sense that it either urges the advisee to do something that is beyond the advisee's control or against the advisee's interest.[2]

It is ironical but should come as no surprise in the light of the above discussion that the pervasive error of the advising economist has been that of futile advising.

In the *Economic Times* of 12 July 1991, Abhijit Sen presents, with his usual clarity, a set of detailed instructions about what the Indian government should do about India's foreign-exchange problem. But having done so, and just as the reader begins to warm to the idea that here at last is the solution to a stubborn problem, Sen goes on to observe, 'But for this [that is, for his advice, to be followed] the existing culture in government must be turned upside down'. But what is the value of advice the prerequisite for which is that government be turned upside down? This is virtually tantamount to saying that the advice cannot be followed. So whatever else be the value of such an essay, as advice it belongs to the category of 'futile'.

In the *National Review* of 30 September 1996, we find Deal Hudson advising America on how to recover its 'intellect and its freedom'. 'Our best bet', he argues, is the 'church-related university, illumined by the light of faith, confident in its curriculum, rooted in history, concerned for the student as a whole person'. In such a university, he goes on to urge, the primary concern should be 'the development of character, the discernment of true values, and the preparation for heaven'. The trouble with this advice is that it seems compellingly beyond the reach of anybody.

Without questioning the content of the advice, one can multiply the examples, given above, of advice that is futile—for which there is no hope of even the most diligent advisee being able to carry it out.

A very different set of problems arises once we move away from traditional models of the economy to ones where one person's utterances can influence another's action. Such a world raises not only issues of analytical interest to the economist or the game theorist[3] but also moral dilemmas for the adviser. And if we are to ever have a science of advising we will need to contend with these issues and dilemmas.

17.3. Advising Endogenous Government

To make room for advice that will not fall on deaf ears, a *necessary* step is to move away from the Walrasian world to one in which information is imperfect and asymmetric. In particular, we shall assume that government consists of individuals (the politicians) with their own aims and objectives but who have inadequate information about the projects or plans from which they have to choose and implement one. On the other hand, there are the advisers who, through training, research, or, for that matter, clairvoyance, have information about the effects of each project or plan. But they are not allowed to choose; they can only advise the politicians about what to choose. In brief, we are taking a small step towards a more realistic model of government by recognizing that (1) government is not an exogenous agent, but a collection of individuals with their own motivations and (2) information in society is incomplete and asymmetric. It is interesting to note that Green's (1993) model, while addressing very different issues, nevertheless uses an argument concerning asymmetric information and communication to explain the emergence of parliamentary democracy.

These assumptions make it possible for one agent to influence another through advice. But (1) and (2) are by no means *sufficient* for this to be so. The aim of this section is to demonstrate this by constructing some simple game theoretic examples.

If the adviser and the advisee have the same objective functions then all is well. There is scope for 'cheap talk' and advice takes place in the usual way, that is, by saying that x should be done when the adviser believes that x should be done (Farrell and Rabin 1996).[4] Trouble arises if the adviser and the politician have different aims.

Let me assume that the adviser works entirely in the interest of the people,[5] while the politician is self-interested; and the people's (and, therefore, the adviser's) interest is not the same as that of the politician.[6] The simplest illustrative example of this (see Basu 1992) is where there are two projects: N (a nuclear power plant) and T (a thermal one), from which the politician has to choose one. Like all such projects the impact of each project on society is very complicated, and the politician does not know what effect the projects will have on his and other people's welfare.

We can formalize this information structure by supposing that there are two states of the world, w_1 and w_2, which occur with probability 1/2 each. In the language of game theory, 'nature' makes an equi-probable choice between w_1 and w_2. In w_1 the pay-offs from N to the adviser and the politician are 1 and 0 and from T the adviser gets 0 and the politician gets 1. In w_2 the pay-offs from N and T are reversed. The adviser knows which state of the world has occurred and makes the first move. He has to choose between saying 'Do N' and 'Do T'. These two actions are denoted by N and T in state w_1, and by N' and T' in state w_2. The politician hears the advice but does not know whether w_1 or w_2 has occurred, and has to choose between the nuclear and thermal power plants. The politician's choice is implemented and the players reap pay-offs as already explained. This game is described using the standard device of a game tree in Figure 17.1, and is called the Orthogonal Game. Note that nodes x and y belong to the same information set, which captures the idea that the politician cannot tell whether he is at x or y, when he is at one of them. In both these nodes he has just heard his rather taciturn adviser say 'Do N'. And, since he does not know whether w_1 or w_2 had occurred, for him x and y are indistinguishable. His choice of the nuclear and thermal plants at these nodes is denoted by n and t.

What will be the outcome of this game? Will the adviser be able to influence the choice of the politician?

Let us first check intuitively how they will play this game. Suppose w_1 occurs.

The adviser will of course want the politician to choose the nuclear project. Let us suppose that he is naive, and so says exactly that: 'Sir, I advise you to go for the nuclear project', or equivalently, 'Do N'. The politician, knowing about her adviser's 'strange' political leanings, would, it seems, promptly choose the thermal plant. This happens in the same way that a child told by the mother not to watch channel 29 that evening because there will be 'a boring film' called *The Last Tango in Paris* knows that that is one evening when the child should not watch *Mr. Rogers' Neighborhood* and instead turn to channel 29.

If the adviser were rational and knew that the politician would do the opposite, then the above outcome would not occur. The adviser may then give

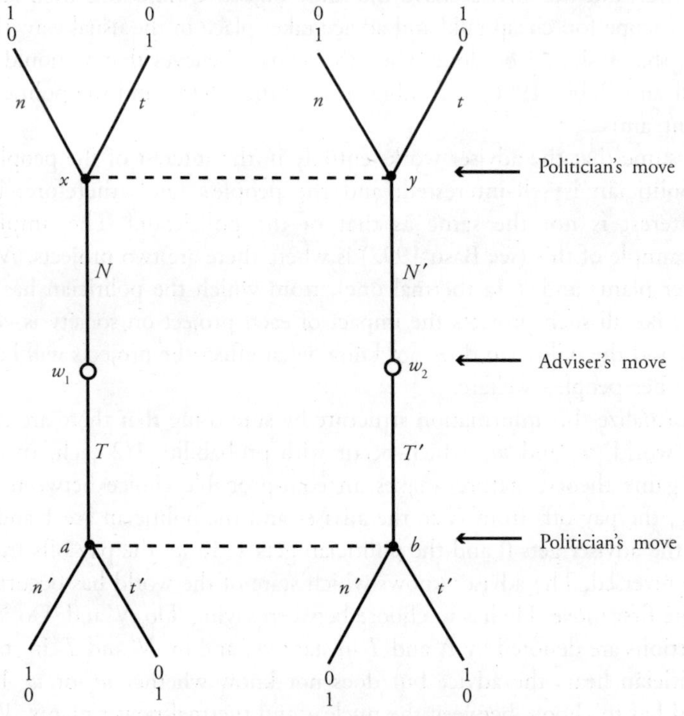

Figure 17.1 The Orthogonal Game

the false advice 'Go Thermal' and hope that the politician will go nuclear. Indeed there are not too many mothers who would instruct their child as in the above paragraph.

But, of course, if the politician knows that the adviser is rational and that the adviser knows that the politician is rational, then this simple trick of the adviser will not work.

It is actually easy to check that the only Nash equilibrium is one in which the adviser's advice is completely uncorrelated to which state of the world actually occurs and the politician's choice is completely uninfluenced by the adviser's advice.

What the above example points to is not just the difficulty of advising but to some deep problems of communication in general. As Loury (1994: 432) points out in his engaging essay on self-censorship and political correctness: 'There is always some uncertainty when ideas and information are exchanged between parties who may not have the same objectives. Each message bears interpretation'. He goes on to point out how he himself (as a prominent commentator on racial issues in the US) has to be cautious (p. 435):

> I must tread carefully as I try to express my particular 'truth'. If you will 'read between the lines' for my true meaning [...], then I am determined to write between the lines—avoiding (or embracing) certain 'code words', choosing carefully my illustrative examples, concealing some of my thinking while exaggerating other sentiments—so as to control the impression I make on my audience.

It may appear that the problem that the Orthogonal Game highlights is the difficulty of advising when preferences are diametrically opposite between the adviser and the advisee. As O'Flaherty and Bhagwati (1996) observe: 'Saddam Hussein is unlikely to revise the Iraqi agricultural price system just because some American economists tell him that doing so would be nice'.[7]

This naturally leads to the suggestion that for an adviser to play a positive role there must be a reasonable affinity of interests between the adviser and the politician and, more generally, the speaker and the listener. Thus O'Flaherty and Bhagwati (1996) speak about the importance of the 'coincidence of interests', and, to stress that this need not be a non-generic special case, add that 'coincidence does not have to be exact'. In the same spirit, Loury (1994: 436) remarks, 'If we know a speaker shares our values, we more readily accept observations from him ...'; and 'when we believe the speaker has goals similar to our own, we are confident that any effort on his part to manipulate us is undertaken to advance ends similar to those we would pursue ourselves'.

What I want to illustrate, however, is that the prognosis is gloomier than these observations suggest. Similarity of objectives is not enough. Anything short of an exact coincidence of preference may result in a complete breakdown in communication. This paradoxical result is driven by a familiar 'infection' argument where a small anomaly or some informational event far away becomes pervasive and has real effects (see Morris and Shin 1995). This is proved in the next section by constructing a game which I call the Cheater's Roulette.

17.4. The Cheater's Roulette

To illustrate the result mentioned in the last paragraph of the previous section, consider a continuum of projects $\Omega = [0, 1]$. What effect each project has on the adviser and the politician is known by the adviser but not by the politician. The adviser gives an advice to the politician which takes the form of saying 'Do x', where $x \in \Omega$ and the politician then chooses some $y \in \Omega$. We shall describe a way of measuring the nearness of the preferences of the politician and her adviser, and show that unless the preferences are identical the politician will not pay any heed to the adviser's advice.

Some abstraction makes it easier to describe this game, which I call the Cheater's Roulette. It consists of a roulette board, the circumference of which is a unit circle. Let us call the northern-most point 0 and the same point also 1 (in the same way that in a clock 0 and 12 refer to the same point). This is illustrated in Figure 17.2. The circumference then is our set of projects $\Omega = [0, 1]$.

The board has a 'hand' which is pivoted to the center of the roulette board. The hand can be made to spin. The line with the arrow in Figure 17.2 denotes the hand. The game is played as follows. The politician sits where she cannot see the board. The hand is given a spin (by 'Nature', let us say) and after it comes to rest the politician is asked to choose a point from Ω. If the hand comes to rest at point m, as shown in the figure, then the politician is paid as follows. She gets 100 dollars if she chooses m; 0 dollars if she chooses the point

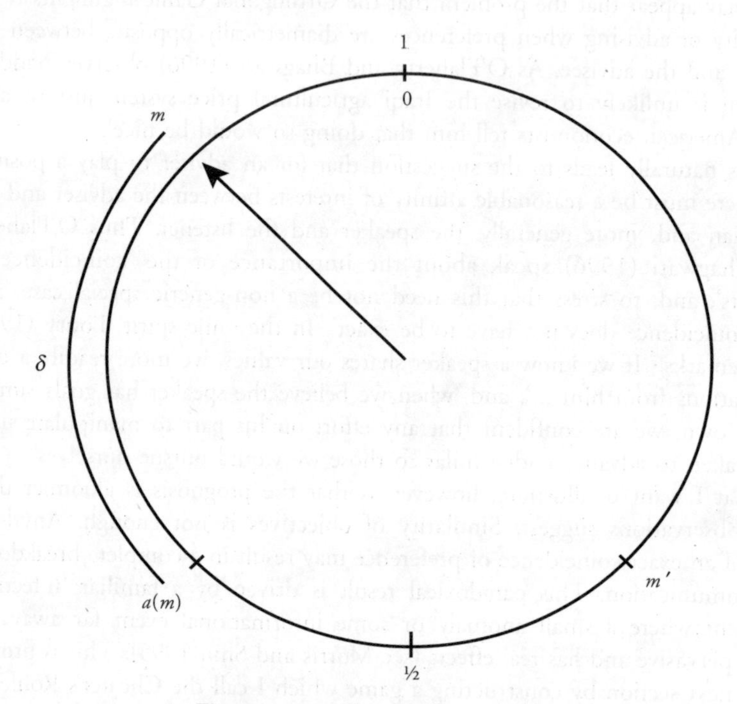

Figure 17.2 The Cheater's roulette

diametrically opposite to m (i.e., point m' in Figure 17.2) and the pay-off drops off linearly (though monotonically would do) as she chooses points further and further away from m.

This may be stated formally as follows. For any two points, x, $y \in \Omega$, the distance between them, denoted by $d(x, y)$ is the shortest distance between x and y along the circumference. It follows that

$$d(x, y) = \min\{|x - y|, 1 - |x - y|\}.$$

If Nature selects $m \in \Omega$ and the politician $x \in \Omega$, then the politician's pay-off is

$$100 - d(x, m)200.$$

Note that $d(x, m)$ can vary between 0 and ½. Hence, the pay-off varies between 100 and 0.

The adviser in this game is actually an accomplice who watches Nature's selection and whispers a piece of advice to the politician about what should choose.

The adviser also gets a pay-off which depends on what Nature and the politician choose. This may be described as follows. Note that whatever Nature chooses is the politician's ideal point. Let us suppose that the adviser is to the 'left' of the politician and define the 'adviser's ideal point', $a(m)$, to be a point which is at a distance of δ ($\leq$ ½) to the left of m. An illustration of $a(m)$, where δ is ¼ is shown in Figure 17.2. To remind ourselves that $a(m)$ depends on δ we could write it as a $a_\delta(m)$ but I am not doing so for reasons of notational simplicity.

If Nature selects m and the politician chooses x, the adviser is paid 100 dollars if $x = a(m)$ (that is, if the politician chooses the adviser's ideal point), 0 dollars if x is diametrically opposite $a(m)$, and the pay-off falls off linearly as x moves away from $a(m)$. Formally, the adviser's pay-off is

$$100 - d(x, a(m))200$$

If $\delta = 0$, then the adviser's preference is exactly the same as the politician's and his every advice will be taken by the latter. The paradoxical result is this: if $\delta > 0$, then no matter how small δ is, communication will break down totally between the adviser and the politician. The only Nash equilibrium is one in which the politician ignores her adviser's whisper in making her choice.[8] Attention is throughout restricted to pure strategies.

In order to sketch a proof of this, I need to introduce some new terminology. Let the adviser's strategy be denoted by ϕ where

$$\phi : \Omega \rightarrow \Omega.$$

For every selection $x \in \Omega$ by Nature, $\phi(x)$ is what the adviser asks the politician to choose.

The politician's strategy is c, where

$$c : \Omega \rightarrow \Omega.$$

For every advice $x \in \Omega$ given by the adviser, $c(x)$ is the element of Ω that the politician chooses.

Hence, if Nature selects $m \in \Omega$, the politician chooses $c(\phi(m))$ and her pay-off is $100 - d(c(\phi(m)), m)200$. And the adviser's pay-off is $100 - d(c(\phi(m)), a(m))200$. Since Nature selects by spinning the hand, it selects from a uniform density function on Ω. Hence, given ϕ and c, we can compute the expected pay-off of each player. Let (ϕ^*, c^*) be a Nash equilibrium of this game. What we need to show is that $c^*(x)$ is independent of x (i.e., the politician's choice is independent of the advice she receives). That is, there exists $y \in \Omega$, such that $c^*(x) = y$, for all $x \in \Omega$.

First note that if $y \in c^*(\Omega)$ then there exist points z and x strictly to the right and left of y, respectively, such that[9] $y \in [x, z]$ and no other point (i.e., other than y) in $[x, z]$ is in $c^*(\Omega)$. If this were not true, we could find $x, z \in \Omega$ such that $c^*(\Omega)$ is dense in the interval $[x, z]$. Then c^* cannot be an optimum strategy for the politician. Let y be in the interior of $[x, z]$, such that for some $r \in \Omega$, $c^*(r) = y$. Then if Nature selects m such that $a(m) = r$, $\phi^*(r)$ must be such that $c^*(\phi^*(r)) = a(m)$. Clearly then the politician would be better off deviating from her choice $c^*(\phi^*(r))$. This establishes the first sentence of this paragraph and thereby proves that $c^*(\Omega) < \infty$.

Denote $c^*(\Omega) \equiv \{x_1, \ldots, x_n\}$ where x_1 is the first point in $c^*(\Omega)$ at or to the right of 0; and $x_2, x_3, \ldots$, follow clockwise as shown in Figure 17.3. Suppose $n \geq 2$.

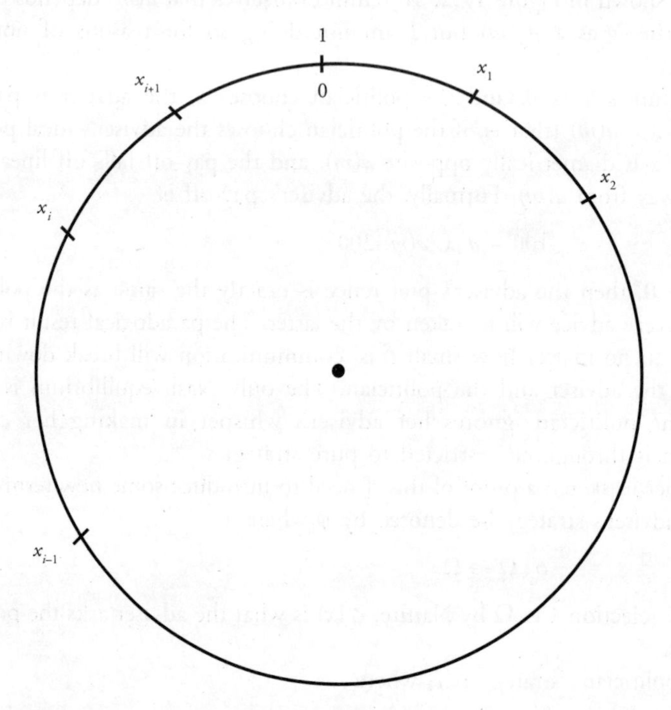

Figure 17.3

It is now easy to check that c^* cannot be optimal for the politicians unless the length of x_i to x_{i-1} exceeds the length of x_{i+1} to x_i (lengths being measured along the anti-clockwise arc). If this is not the case it is possible to check that the politician can do better by choosing slightly clockwise away from x_i when the advice y is such that $c^*(y) = x_i$. Since the projects belong to a modular number system, it is not possible for the gaps between adjacent x_is to increase throughout as we move in one direction. Hence, $n = 1$. *This completes the proof of the paradoxical result.*

What we have proved is this. There exists $x \in \Omega$, such that for all $y \in \Omega$, $c^*(y) = x$. In equilibrium the adviser's advice has no effect on the politician's choice. The politician just arbitrarily picks a point $x \in \Omega$. It follows that in equilibrium the adviser just babbles—he gives advice which conveys no information about the state that has occurred (that is, Nature's choice).

Before proceeding further it may be useful to explain my use of a unit circle, as opposed to the more conventional unit interval. The immediate reason for doing this is the mathematical convenience of being able to describe one person as being a distance δ to the left of another, irrespective of the latter's location. Second, this description can describe some very real problems. Some years ago, the Ministry of Finance in Delhi was considering changing the time of the annual budget of the Government of India (currently the budget year starts April 1). There were many real issues involved. A budget in October, for instance, would mean that we would know how the monsoons have been and, therefore, be better placed to plan ahead. There were in fact so many variables involved that expert advice was sought about when the budget year should start. If we now think of the unit circle as representing the calendar year from 1 January to 31 December, then this problem has the same algebra as our model.

Can advice then have no role unless there is a total coincidence of preference? Despite the above result the answer to this question is in the negative. First, there is now a small body of literature that shows that even when precise advising can be shown to have no role, ambiguity in speech or vagueness of expression can be used to convey *some* information from the speaker to the listener (see the seminal work of Crawford and Sobel 1982; also Stein 1989). This in itself is a very interesting result because it shows how moving away from precision to ambiguity may help us to actually convey more.[10]

Second, we could appeal precisely to the fact that the outcome of the cheater's roulette is *paradoxical* and urge the reader to reject the game-theoretic solution. In this respect the cheater's roulette is akin to the traveller's dilemma (Basu 1994) because in that game (in a sense) it is rational to reject playing the game rationally because it seems reasonable to expect that the other player will do the same. Now put yourself in the shoes of the adviser in the cheater's roulette and suppose that δ is very small, that is, the adviser and the advisee have almost identical objectives. If the roulette hand stops at m, one reasonable way of thinking is this: 'A Nash equilibrium play could make both of us lose a lot. So why don't I advise something in the vicinity of m. Surely the politician will also realize that the Nash equilibrium play does us no good and so choose something in the vicinity of *m*'. This is not watertight reasoning but nevertheless not one to be dismissed. Note also that 'the vicinity of *m*' is an imprecise expression. But, as I had argued in Basu (1994), meta-rational behaviour depends on the use of imprecise (and hence realistic) categories of thinking.

No matter how we seek to resolve the problem, one thing is evident. Barring the non-generic special case in which there is a complete coincidence of wants, it does not pay to give the advice that one believes in. That is, the same morality that leads you to think that 'X should be done' prompts you not to *say* 'X should be done'. This creates a moral dilemma which may have no easy solution and this is the subject matter of the next section. But before moving on to it I want to dwell on two caveats of the present analysis and a related observation.

First, we have in this chapter, for reasons of tractability, taken 'advice' to be simple normative statements or, more generally, any message which can influence behaviour by *informing* the advisee. However, in reality advice often takes the form of persuasion, which involves attempts to influence the advisee's *preference.* What complicates this is the fact that people often voluntarily go for this kind of advice. This is true not just of the alcoholic seeking counselling, but in politics and in government there is the continuous play of forces jockeying and buffeting to influence preferences and of individuals voluntarily leaving themselves vulnerable to preference shifts.

Second, once we move away from the assumption of preferences being mutually known, communication (and, for that matter, certain kinds of actions) may acquire some new strategic element whereby the communicator seeks to influence the belief of the listener about the communicator's preference. Preferences of political actors are often important to others because they indicate what the politician might do in contingencies which arise in the future (some of which may not even currently be conceivable). Thus it is arguable that President Clinton supported the Helms-Burton act aiming to punish companies outside the US that do business with Cuba not because he believes in it (there is actually some evidence suggesting that he does not), but because he wanted to *appear* more conservative before the presidential election and thereby create the expectation that in future decision making he will pick the more conservative alternative.

By assuming that preferences are exogenously given and common knowledge among the agents, the present Chapter stays clear of these two complications. But they are important in reality and deserve to be on the agenda for future research.

Finally, the related observation. I have posed the problem of communication in the asymmetric context involving an adviser and an advisee. The same problem can, however, also arise when the agents involved are symmetrically placed, for example the members of parliament taking a vote or the members of a jury deciding by majority. An interesting paper by Austen-Smith and Banks (1996) illustrates how the dilemma of not revealing one's information sincerely can arise in the context of problems such as the celebrated Condorcet jury theorem. In their model this can happen even when all members have the same preference.

17.5. A Moral Conundrum

An act of speaking or writing usually has consequences for the world. The *Communist Manifesto* was nothing but some words on paper. So were the Bible, the Koran and even *The Satanic Verses.* But these 'words' have had consequences for the world of action, creating or destroying wealth, stirring human beings into acts of bravery or cowardice. Hence, anyone who writes or speaks has to take into account the consequences of the writing and the speech. This is especially so for those who write for and speak to large audiences. Nevertheless, the scientist writing *positive* science can at least claim no *inconsistency* if he chooses to write whatever is the truth with no thought to the consequence of his writing.

On the other hand, a scientist, or, for that matter, anybody, making a normative statement may face a deeper moral conundrum, a problem which is virtually one of consistency. The politician, beseeching the public to act, the economist advising the politician, and the journalist urging the economist to say something, all face this problem. Unlike the positive scientist, these people cannot disclaim having a normative purpose because their very act of speaking reveals it.

As we have already seen, when an expert or an informed person utters something, people try (or should try) to elicit information from that utterance, in the same way that the politician in the cheater's roulette tries to deduce the outcome on the roulette board from his adviser's whisper. Similarly when you read or hear Mr Stephen R. Covey, the author of popular improve-yourself books such as *The Seven Habits of Highly Effective People,* tell you to be goal oriented or to keep 'the end in mind' you have reason to believe that *he* believes in being goal oriented. Given that it seems unlikely that Mr Covey's goal is to make *you* goal oriented; and his much more likely goal is to maximize the sales of his books, you have reason to suspect that some of the things he advises people are the advice people *like* to hear. And you could decide, not totally unreasonably, that, given the great success of his books, following not his advice but him is the more profitable strategy. Of course, you may be wrong in attributing the best-seller motive to Mr Covey; he perhaps has a missionary purpose. But the fact remains that people do look for meanings other than the one explicitly stated,[11] which is the subject matter of Kuran's (1995) persuasive book.

Now suppose you want to tell government: 'Government should do *x*'. This will make people try to guess what you know and they do not, make them act in certain ways and, let us suppose, bring about the kind of world that you do not morally approve of. And suppose your giving the opposite advice will bring about a desirable world (in terms of your own morals). Then you face a moral conundrum because what is in conflict is not your self-interest with your moral judgement but your morals with your morals. Should you say what you believe in or should you say the reverse and bring about the kind of world that you believe in? Note that a person making a normative statement cannot even use the alibi of being normatively disinterested. He has to confront the dilemma.

In brief, this is a conundrum that one has to confront if one wishes to advise and pronounce publicly on policy. It may be possible to construct models of repeated advice which bring the two moral options discussed in the above paragraph into alignment. That is, it may be morally best to say what you actually believe in because otherwise your 'strategic' behaviour will get revealed in the long run. But at this stage we have no option but to leave this problem as an open-ended issue since even if there are repeated-game stories which can resolve it, these are not transparent. Till this is resolved we will be right in trying to read between the lines of not only what politicians and other government officials say, but also what the economic adviser and the economist in public life say.

17.6. Remarks on Endogenizing Government

It is natural for the modern person to take government for granted. Government is a necessary concomitant of state and, for certain discourses, it is state. Yet in the history of human beings, state and government are relatively modern institutions. People belonged to tribes and had chiefs rather than governments and heads of states. This is true even in some contemporary cases. Many tribals, for instance those living in the Andaman and Nicobar Islands, are not aware that they are Indians. To them the agents of the state—the police, the civil servants—are not representatives of the 'law' but, on the contrary, illegal trespassers on whom the use of poison arrows is considered well worth the poison. Barring some such small exceptions, all people treat government and its agencies as part of life.

As Strayer's (1970) elegant little book reminds us, this was not always so. Up to as late as the eleventh century AD there were no states as we know now. There were some small city states and there were empires. According to Strayer, state as a moral authority, as an agent with a 'monopoly of the legitimate use of physical force'—to steal a description from Max Weber—and as an institution for providing public goods is a phenomenon of the last millennium.

Why did the state emerge in the twelfth and thirteenth centuries? This is a subject matter for history and one that may well require the longevity of a historian's professional life for successful investigation, but one suggestion in Strayer's essay is particularly interesting. The twelfth and thirteenth centuries saw a steep rise in learning and literacy. It therefore allowed the codification of law and the signing of contracts in a way that may not have been possible earlier. This gave rise to the need for an enforcer of contracts and government soon became that ubiquitous 'third party', the enforcer of contracts. Modern society, it is arguable, would be unsustainable without an institution for supporting contracts and covenants. The prosperity of contemporary economies owes as much to the slow evolution of this institution as it does to the many sudden scientific breakthroughs.[12]

It is now increasing recognized that the 'market' cannot function efficiently unless it is 'embedded' in suitable institutions (Granovetter 1985; Platteau 1994).[13] For reasons of *analytical* convenience, economists often ignore this embedding feature. What is troubling is that, having made this assumption of treating the market as functioning in a vacuum, many economists forget that this was an *assumption.* This can be costly.

Let me illustrate this with an example. One of the two celebrated 'fundamental theorems' of welfare economics may be stated, taking the liberty of colloquialism, as follows:

> If individuals maximize their own selfish utility, then (given that certain technical conditions are satisfied) the competitive equilibrium that arises is always Pareto optimal.

In itself, this is a mathematical theorem which tells us nothing about the real world. Its application to the real world depends on how we interpret it, *beyond* what it actually tells us. One popular interpretation of this theorem is that it shows that if individuals are left free to choose whatever they want to, then society attains optimality. And, conversely, taxes and other government interventions which limit the feasible sets of individuals tend to result in suboptimality. In defence of this position one would typically point out that in the standard competitive model in which the theorem is established, it is assumed that individual consumers are free to choose any point (or basket of goods) from within their budget sets (or what their incomes permit).

Let us now think for a moment what are the kinds of choices a person can in reality make. It is true that a person can choose from a variety of alternative baskets of goods which lie within his budget; but a person can also choose to rob, steal, and plunder; he can try to take away the endowment of another individual, invoking the age-old principle of more being better than less; he can commit blackmail, larceny, and arson. Hence, when we allow an individual to choose any point within his budget set, there are two ways of viewing this. We could view this as giving him great freedom: he can choose *any* point; or view this as very restrictive: he is not allowed to choose anything (from the large menu of options he has in life) apart from choosing a point from his budget set. If we follow the latter, then here is a view of the fundamental theorem which is a 'dual' to the popular interpretation: The fundamental theorem shows that society attains optimality if individual choice is severely restricted and, in particular, confined to choosing points from within the budget set.

One may argue that the restriction of not allowing theft of other people's endowment, larceny, and blackmail is not a restriction at all but is in the self-interest of the individual. If that is so then we need to formally show this by starting with a model where all these 'extra-economic' activities are allowed. And once we start from such a large domain, to get to the case where the pure general-equilibrium model works, we need the institution of government or some other related institution[14] to prevent individuals from finding some of these 'extra-economic' activities worthwhile. Since this exercise or embedding the market model has not been done thus far, we do not really know whether the model of the market, abstracted from its social and political moorings, can ever be realized. This would be an easy agenda if government were conceived of as an exogenous body that makes it costly for individuals to steal and rob. But as argued through this chapter, this is not a permissible strategy. We have to allow for thc fact that government is run by individuals, who respond to incentives, and explain the survival of government and government's power from a model of individual decision making.

In order to model the survival of government and the people's respect of authority, we need to make room for 'triadic' arrangements (Basu 1986) in our economic models. We need to allow for the fact that *i* respects government authority for fear of what *j* will do to *i* (for instance ostracize *i*) if *i* shows open disrespect towards governmental authority. In Friedman's (1994: 10) words: 'I will accept one [the tax collector] and fight the other [the robber] because of my beliefs about other people's behaviour—what they will or will not fight for. We are bound together by a set of mutually reinforcing strategic expectations'.

Once we have modelled government as an endogenous part of the economy, we can no longer will government into action whenever we wish and however we wish, nor can we treat it as the repository of exogenous variables for our macromodels; but we shall have a description of an economy that reflects reality much better, and, though the policy advisers will probably have to practise greater reticence than they currently do, they will know better *how* to give policy advice.

NOTES

1. Magee and Brock (1983); Barro (1984); Friedman (1986); Basu (1992); Srinivasan (1992); Austen-Smith (1990). The subject of endogenizing government policy and policy change in order to gain a better understanding of development is addressed in Ranis and Fei (1988).
2. A more detailed taxonomy of 'bad advice' is developed in Basu (1992).
3. The rapidly growing literature on cheap talk is testimony to this (see Crawford and Sobel 1982; Stein 1989; and, for a recent survey, Farrell and Rabin 1996).
4. Even here there can be trouble if there is co-ordination of action needed between the adviser and the advisee and there is a non-zero probability of an advice failing to be common knowledge after it is announced (Rubinstein 1989; Halpern and Moses 1990).
5. Nothing formal hinges on this assumption but it prepares the ground for some moral dilemmas that are discussed in the next section.
6. It is important to understand that the fact that one may try to delude one's listeners does not in itself suggest a selfish motivation. As Goffman (1959: 18) points out: 'It is not assumed, of course, that all cynical performers are interested in deluding their audiences for purposes of what is called "self-interest" or private gain. A cynical

individual may delude his audience for what he considers to be their own good, or for the good of the community'.

7. And, as the orthogonal game shows, neither should Saddam Hussein not revise it *because* he has been asked to revise. Such a response would also make him vulnerable to manipulation.
8. Some reader may wish to skip the proof of this claim which stretches over the next few paragraphs up to the point where it says. '*This completes the proof of the paradoxical result*'.
9. For $x, y \in \Omega$, $[x, y]$ denotes the shorter arc between x and y, the tie being broken arbitrarily for x and y which are diametrically opposite to each other.
10. There is a related (though analytically distinct) result which shows how an incumbent politician confronting an ill-informed electorate may gain most by being ambiguous about his preference (Alesina and Cukierman 1990).
11. And, while on the topic of popular writing, one may quote the celebrated Agony Aunt, Ann Landers, warning people not to be too literal: 'Resist the temptation to tell your friends about your indigestion. 'How are you?' is a greeting, not a question' (*The Ithaca Journal,* 5 October 1996).
12. In understanding the state, an alternative to studying its origins is to examine the conditions for its existence. 'Functional theories of the state', as these are called in the political science literature (Tilly 1975), study the concomitants of the national state. Like the market, the state also is in reality embedded in other institutions. An understanding of the latter can give us insights into the very meaning of the national state.
13. Bowles and Gintis (1992) try to show that the Walrasian model instead of capturing the consequences of perfectly rational, self-interested behaviour by individuals, describes a situation where individual rationality is restricted to certain domains of decision making.
14. A range of alternative institutions that can emerge out of individual initiative to help co-operation is discussed in Ostrom (1990). The position that I am taking is that government is also one such institution.

References

Alesina, A. and A. Cukierman. 1990. 'The Politics of Ambiguity'. *Quarterly Journal of Economics* 105: 829–50.

Anderson, T.J. and P.J. Hill. 1975. 'The Evolution of Property Rights: A Study of the American West'. *Journal of Law and Economics* 18: 163–79.

Austen-Smith, D. 1990. 'Information Transmission in Debate'. *American Journal of Political Science* 34: 124–52.

—— and J. Banks. 1996. 'Information Aggregation, Rationality, and the Condorcet Jury Theorem'. *American Political Science Review* 90: 34–45.

Barro, R.J. 1984. 'Discussion [of Sargent's paper]'. *American Economic Review* 74 (May): 179–87.

Basu, K. 1986. 'One Kind of Power'. *Oxford Economic Papers* 38: 259–82.

—— 1992. 'Bad Advice'. *Economic and Political Weekly* 27: 525–30.

—— 1994. 'The Traveler's Dilemma: Paradoxes of Rationality in Game Theory'. *American Economic Review* 48 (May): 391–5.

Bhagwati, J. 1990. 'The Theory of Political Economy, Economic Policy, and Foreign Investment'. In M. Scott and D. Lal, eds. *Public Policy and Economic Development.* Oxford: Clarendon Press.

——, R. Brecher, and T.N. Srinivasan. 1984. 'DUP Activities and Economic Theory'. In D. Colander, ed. *Neoclassical Political Economy.* Cambridge, MA.: Ballinger.

Bowles, S. and H. Gintis. 1992. 'Power and Wealth in a Competitive Capitalist Economy'. *Philosophy and Public Affairs* 21: 324–53.

Buchanan, J.M. 1968. 'An Economist's Approach to "Scientific Politics"'. In M. Parsons, ed. *Perspectives in the Study of Politics.* Chicago: Rand McNally.

Crawford, V. and J. Sobel. 1982. 'Strategic Information Transmission'. *Econometrica* 50: 1431–51.

Dixit, A. 1996. *The Making of Economic policy.* Cambridge, MA.: MIT Press.

Farrell, J. and M. Rabin. 1996. 'Cheap Talk'. *Journal of Economic Perspectives* 10: 103–18.

Friedman, D. 1994. 'A Positive Account of Property Rights'. In E.F. Paul, F.D. Miller, and J. Paul, eds. *Property Rights.* Cambridge: Cambridge University Press.

Friedman, M. 1986. 'Economists and Economic Policy'. *Economic Inquiry*: 1–10.

Goffman, E. 1959. *The Presentation of Self in Everyday Life.* New York: Doubleday.

Granovetter, M. 1985. 'Economic Action and Social Structure: The Problem of Embeddednes'. *American Journal of Sociology* 91: 481–510.

Green, E.J. 1993. 'On the Emergence of Parliamentary Government: The Role of Private Information'. *Federal Reserve Bank of Minneapolis Quarterly Review*: 2–16.

Halpern, J.Y. and Y. Moses. 1990. 'Knowledge and Common Knowledge in a Distributed Environment'. *Journal of the ACM* 3: 549–87.

Hobbes, T. 1651. *Leviathan.* ed., R. Tuck. Cambridge: Cambridge University Press.

Kuran, T. 1995. *Private Truths, Public Lies: The Social Consequences of Preference Falsification.* Cambridge, MA.: Harvard University Press.

Loury, G. 1994. 'Self-Censorship in Public Discourse: A Theory of 'Political Correctness' and Related Phenomena'. *Rationality and Society* 6: 428–61.

Magee, S.P. and W.A. Brock. 1983. 'A Model of Politics, Tariffs and Rent-seeking in General Equilibrium'. In B. Weisbrod and H. Hughes, eds. *Human Resources, Employment and Development* 3. London: Macmillan.

Morris, S. and H.S. Shin. 1995. 'Informational Events which Trigger Currency Attacks'. University of Pennsylvania. Mimeo.

O'Flaherty, B. and J. Bhagwati. 1996. 'Will Free Trade with Political Science Put Normative Economists Out of Work?' Columbia University. Mimeo.

Ostrom, E. 1990. *Governing the Commons: The Evolution of Institutions for Collective Action.* Cambridge: Cambridge University Press.

Platteau, J.P. 1994. 'Behind the Market Stage where Real Societies Exist: The Role of Public and Private Order Institutions'. *Journal of Development Studies* 30: 533–77.

Posner, R.A. 1981. *The Economics of Justice.* Cambridge, MA.: Harvard University Press.

Ranis, G. and J.C.H. Fei. 1988. 'Development Economics: What Next?' In G. Ranis and T.P. Schultz, eds. *The State of Development Economics: Progress and Perspectives.* Oxford: Blackwell Publishers.

Rubinstein, A. 1989. 'The Electronic Mail Game: Strategic Behaviour under Complete Uncertainty'. *American Economic Review* 79: 385–91.

Stein, J.C. 1989. 'Cheap Talk and the Fed: A Theory of Imprecise Policy Announcements'. *American Economic Review* 79: 32–42.

Srinivasan, T.N. 1992. 'Bad Advice: A Comment'. *Economic and Political Weekly* 27: 1507–8.

Strayer, J.R. 1970. *The Medieval Origins of the Modern State*. Princeton: Princeton University Press.

Taylor, M. 1976. *Anarchy and Cooperation*. London: John Wiley.

Tilly, C. 1975. 'Western State-Making and Theories of Political Transformation'. In C. Tilly, ed. *The Formation of National States in Western Europe*. Princeton: Princeton University Press.

18 The Economics of Tenancy Rent Control

with Patrick M. Emerson

In early 1996, when New York City's rent-control law came up for evaluation and possible modification, the public debate spilled over beyond New York to national newspapers and the international media. The same questions that arose in this debate have arisen in the past in discussions concerning rent control in, among other places, France, Germany, India, Sweden, and other parts of the United States. These debates reveal, more than anything else, how widely the central issues of rent control are misunderstood. Part of the blame for the popular misunderstanding of the effects of rent control lies with economists. Despite quite a substantial literature on the subject, some of the key analytical questions, especially ones concerning the relation between inflation and rental adjustment, remain unanswered.

The aim of the present chapter is to construct a model that captures the main stylized features of a form of rent control pervasive around the world. We refer to it as 'tenancy rent control', Tenancy rent control, which is a special case of what is known in the literature as 'second-generation rent control', allows landlords to choose a nominal rent freely when taking on a new tenant (the tenant is of course free to reject the offer) but places restriction on raising rents on, or evicting, a sitting tenant. This causes an erosion in the real value of rent if a tenant stays on for too long, whenever there is positive inflation in the economy, which for most economies is true most of the time. This means that landlords will prefer short-staying tenants to long-staying tenants. Since a tenant's type will be better known to the tenant than the landlord, the tenancy

From *The Economic Journal* 110 (October), 2000: 939–62.

The authors have benefited from presentations of this paper at the American Real Estate and Urban Economics Association Annual Conference and at the Applied Microeconomics Workshop at Cornell University. For comments and discussion, we would like to thank Richard Arnott, Pinaki Bose, Franz Hubert, Robert Masson, Ted O'Donoghue, Edgar Olsen, and Buhong Zheng market will be characterized by asymmetric information. Our basic model describes the tenancy market as a model of asymmetric information in which the tenants' types are exogenously given. It is shown that the presence of tenancy rent control will, in general, result in a Pareto sub-optimal equilibrium, whereas a system of free contract will be Pareto optimal. Of course, this does not mean that moving from the former to the latter would make everybody better off. However, the model does illustrate how the real conflict of interest is not between landlords and tenants, as portrayed in most popular debates on rent control, but between tenants of different kinds. This is a result that will come as no surprise to economists. In our case this basic model and result serve as a benchmark that can thereafter be developed to obtain some surprising results.

After constructing the basic model we develop it by endogenizing the tenant types. That is, we allow for the fact that the outcome in the rental market may affect the tenant's lifestyle, for instance, discouraging him from shifting too many times. Once the tenant's 'type' is modelled as an endogenous variable we get the surprising result that rent control may give rise to multiple equilibria. This is a very natural result, based on weak and realistic assumptions, but it seems to have gone unnoticed in the literature. If the economy gets locked in the 'bad equilibrium', among the many possible equilibria, the removal of rent control cannot only bring about an efficient outcome but cause an across-the-board lowering of rents, thereby leaving all tenants better off. This result is established in Section 18.4.

We should clarify that all our comparisons of different rent-control regimes take the form of comparative statics. We do not consider switch-overs from one regime to another. Hence all of the policy prescriptions that flow out of this exercise concern new tenants and new contracts. We do not comment on how, or for that matter whether, changes should be made to laws applicable to currently sitting tenants.

The next section is about the institution of rent control. It discusses different kinds of rent control, some stylized facts, and the real-world context of our theoretical constructions.

18.1. The Institution of Rent Control

In the United States and Europe, the numerous governmental controls in the rental housing market, which are generally described as 'rent control', arose during World War II in response to the mass disruptions caused by the war. After the war, New York City and many European jurisdictions retained versions of these out of fear that the return of troops would send rents skyward. In other parts of the United States, the social upheaval and high inflation of the 1970s was a driving force behind the re-implementation of rent controls. California, Connecticut, Massachusetts, New Jersey, and New York are all states where jurisdictions implemented rent-control policies during this period. While a number of jurisdictions have since abandoned or relaxed rent-control laws, these laws are still commonly found in the United States and the world over.[1]

Before we proceed to discuss rent control, it is worthwhile clarifying that the *absence* of rent control can be of two kinds: the kind with no government intervention in the rental housing market, or the kind where the government allows and enforces contracts (subject to the standard restrictions on the freedom of contract provided under contract law). In this chapter, when we consider a regime with no rent control, we shall be concerned with the latter, which will be referred to as a 'free contract' regime. By its converse, 'rent control' is a generic term that describes rental laws, which place additional restrictions on allowable contracts.

While many different forms of rent control exist in the world, we will focus on a stylized version, which is widely used throughout the world. We will focus on a rent control regime that does not allow the eviction of a sitting tenant and that limits the amount a landlord may increase the rent on a sitting tenant (enough so that rents do not, typically, keep up with inflation). Upon vacancy, however, the landlord is free to negotiate a new rent with a new tenant. This is precisely the regime that exists in quite a few US communities, including Los Angeles, Berkeley, Santa Monica and Palm Springs, and is similar to the system that exists in Washington, D.C. (Dreier 1997; Olsen 1990).[2] This is also a good approximation of rent-control laws in other communities in the United States and elsewhere in the world including Germany, France (Hubert 1995; Börsch-Supan 1986), Sweden, and virtually all major cities in India.

In Delhi, for example, Section 6 of the *Delhi Rent Control Act,* 1958, allowed a maximum of a 10 per cent rent hike every three years, no matter what the inflation. In India the average inflation every three years has exceeded 20 per cent. The Act also made it virtually impossible to evict a tenant. The 1958 Act has subsequently been superseded by the *Delhi Rent Act,* 1995, which is only slightly more flexible.

In New York City, properties under 'rent stabilization' are closest to the rent-control law just described. There have been two major rent-control regimes in New York, 'Rent Control' and 'Rent Stabilization'. Rent Control was a strict regime started in 1947 that assigned rents for individual properties and allowed minor increases. This policy currently covers slightly more than 70,000 units in New York City and is declining with vacancy decontrol and shifts to stabilization. Much more common are properties under Rent Stabilization. This system was implemented much later, in 1969, and was a less stringent form of rent regulation where periodic rent increases are allowed as approved by a rent regulatory board. In 1971, a policy of vacancy decontrol was instituted for all units under either Rent Control or Rent Stabilization after they had been 'voluntarily' vacated. However, in 1974 an amended policy was introduced that ended vacancy decontrol of stabilized units and controlled units in buildings with more than five units in total. The new policy kept the units under stabilization, but allowed increases of between 14 and 16 per cent above the most recently approved allowable rent increase for a voluntarily vacated apartment.[3] Recently, the 1997 agreement to renew New York's Rent Stabilization laws upped the allowable vacancy increase to 20 per cent, for a two-year lease, and an additional 0.6 per cent for each year the previous tenant had occupied the unit.[4] There are other ways for landlords to raise the controlled rent over and above these increases as well. One way is through pass-through costs. This allows landlords who spent money on improvements to a rent stabilized apartment to raise the rent by appealing to the board. Another way is by pleading hardship or increased operating costs. Anecdotal evidence suggests that landlords raise rents through these channels often, even if the amount of money spent on improvements is small. As both of these tactics are sure to meet opposition from a sitting tenant, the common practice is for the landlord to appeal for these increases upon vacancy.[5]

New York City's rent control laws also provided non-rent protection for tenants. In particular, the Rent Stabilization Code stipulated that landlords must offer tenants a renewal lease (at the stabilized rent) before the expiration of the current lease. It also limited the set of circumstances in which the landlord could evict a sitting tenant (non-payment of rent, for example).[6] These provisions are essentially the non-eviction of a sitting tenant clause that we stipulate for the stylized rent-control law studied in this chapter. Thus, the current situation in New York City is very close to the tenancy rent-control regime of our model. Therefore, it seems to us, that tenancy rent control is pervasive, especially when one recognizes that a rent-control system applied to a unit rather than the tenancy, in which rent increases are permitted unless appealed against by the tenant, is not very different.

There has been a considerable amount of theoretical work on rent control. The textbook version of rent control is a price-ceiling model of supply and demand that relates most closely to what Arnott (1995) describes as 'first generation rent control'. These are akin to the rent-control regime which New York City implemented in 1947, in which rents were fixed at a level and rarely allowed to rise. The textbook model has been advanced in several directions (see, for instance, Hubert 1996; Raymon 1983; Frankena 1975; Sweeney 1974; and for a related surveyee Smith et al. 1988). There has also been work done on the political economy of rent control (see Epple 1998; and Fallis, 1988), which tries to explain why the rental laws are what they are by looking at the power structure of different lobbies.

Our aim is, however, more limited—to study the *economics* of tenancy rent control in economies with positive inflation.

18.2. The Basic Model

Let us assume that there are n types of potential tenants in an economy. If N is the set of types, then $N = \{1, \ldots, n\}$. Suppose a fraction p_i of all tenants are of type i. Thus $p_1 + \ldots + p_n = 1$. All agents in this chapter are infinitely lived.

A tenant's type basically refers to how long a tenant stays in the same apartment before moving to a new one. Let t_i be the number of months a tenant of type i stays in the same apartment. Without loss of generality we assume that

$$t_1 < t_2 < \ldots < t_n.$$

In other words, type 1 tenants are the restless souls. Either they have a preference for quick change or have transferable jobs. Type n tenants are the types who gather moss. Others are somewhere in between those extremes. Of course, in reality, depending on the rent-control regime that prevails in an economy, a person may decide to quit a transferable job and take up a stable job or vice versa. But we will, for now, assume that the tenant types are given. This assumption is relaxed in the next section.

Throughout this chapter we assume that there is a finite number of tenants and a finite (and therefore discrete) number of tenant types; and that rents are paid at discrete time intervals. But our method and all the results in this section extend easily to the case where there is a continuum of tenant types and rents are paid continuously.

This is a model with asymmetric information. Each tenant knows his type but a landlord cannot tell the tenant's type by looking at him. In addition, our rent-control law does not allow quit-contingent contracts, rent escalation clauses for long stayers, nor length-contingent payments to tenants. The monthly rent has to be fixed at the time of taking on a new tenant and may not be adjusted for the duration of the tenancy.

Note that even though a tenant's type is unknown to the landlord at the time the tenant moves in, the tenant's type gets revealed at the time the tenant moves out. Hence, by charging a lump-sum amount at the time of a tenant's moving out, a landlord can overcome the problem of asymmetric information. A rent-control law typically prevents such complicated contracts and thus causes the asymmetric information problem to persist (Basu 1989), in fact without such restrictions a rent-control law would be rendered ineffectual. Initial deposits such as key money (or what in India is called *pugree*) with agreement to return a part of it depending on when the tenant leaves, or any kind of rent-escalation clause may be viewed as the market's way to get around rent control. In our model we assume that these types of payments and clauses are not allowed by the law. In other words we are about to analyse the case of 'tenancy rent control', as described in Section 18.1.

We will also assume that there is inflation in this economy which erodes the value of money each month by $1 - \beta$, which is greater than zero. That is, we are assuming—what is nearly universal—that there is some inflation in the economy.

Let the discount factor for all individuals be $\delta \in (0, 1)$, for each month.

If a landlord charges a rent of \$1 per month in real terms from a new tenant and somehow gets only tenants of type i, then the stream of income earned by the landlord, in real terms, is given by

$$1 \;\beta\; \beta^2 \ldots \beta^{t-1} \;1\; \beta\; \beta^2 \ldots \beta^{t-1}\; 1 \ldots$$

Given the presence of rent control, this stream of income is easy to understand. Since the market rent for a *new* tenant is 1, the landlord earns 1 in period 1. Since the rent-control law does not allow the nominal rent to be changed, and the inflation rate is $1 - \beta$, in the second period (third period) the \$1 is equal to β dollars (β^2 dollars) in real terms. This explains the second and third terms in the stream shown above. After t periods the tenant quits. The new tenant pays a rent of \$1 dollar in real terms (or β^{-t} dollars in nominal terms). This explains why the tth term is 1 and so on. The present value of the above stream using the discount factor of δ is denoted by v_i and this is given by:

$$v_i = 1 + \beta\delta + (\beta\delta)^2 + \ldots + (\beta\delta)^{t_i - 1} + \delta^{t_i} v_i \tag{18.1}$$

or,

$$v_i = \frac{1 - (\beta\delta)^{t_i}}{(1 - \beta\delta)(1 - \delta^{t_i})}. \tag{18.2}$$

LEMMA 18.1. If $i < j$, then $v_i > v_j$.

The proofs of this and of all following lemmas (save Lemma 18.3 for which an intuitive proof is given in the text) are found in Appendix 18.1.

Continuing with the case in which rent is \$ 1 per month, let us denote $v_{(i)}$ as the expected present value of returns to the landlord when all tenants of type i or above make themselves available to the landlord as potential tenants from whom the landlord randomly selects one.[7] Then, clearly $v_{(n)} = v_n$. And, more generally,

$$v_{(i)} = \sum_{k=i}^{n} \left(\frac{p_k}{\sum_{j=i}^{n} p_j} \right) [1 + \beta\delta + (\beta\delta)^2 + \ldots + (\beta\delta)^{t_k - 1} + \delta^{t_k} v_{(i)}] \tag{18.3}$$

or

$$v_{(i)} = \frac{\sum_{k=i}^{n} \left(\frac{p_k}{\sum_{j=i}^{n} p_j} \right) [1 + \beta\delta + (\beta\delta)^2 + \ldots + (\beta\delta)^{t-1}]}{1 - \sum_{k=i}^{n} \left(\frac{p_k}{\sum_{j=i}^{n} p_j} \right) \delta^{t_k}}. \tag{18.4}$$

Since the above expressions are worked out assuming that the rent is 1\$ (in real terms for a new tenant), it is now easy to work out the expressions for the case when the rent is \$$R$. If the landlord gets only tenants of type i, we denote the present value of her rental income as $\tilde{v}_i(R)$ and clearly

$$\tilde{v}_i(R) = Rv_i \tag{18.5}$$

where v_i is given by (18.2).

If the rent is R and only tenants of type i and higher seek tenancy, we denote the landlord's present value of income from leasing out one apartment by $\tilde{v}_{(i)}(R)$. Clearly,

$$\tilde{v}_{(i)}(R) = Rv_{(i)} \tag{18.6}$$

where $v_{(i)}$ is given by (18.4).

LEMMA 18.2. If $i < j$, then $v_{(i)} > v_{(j)}$.

As we discussed in the introduction, one of the most popular variants of rent control takes the form of disallowing landlords from raising rents adequately or evicting tenants. Let us, in particular, assume that a landlord can choose a rent, R, at the time of taking on a new tenant (who of course has the freedom to turn down the offer); but then the rent remains (nominally) the same as long as the tenant stays on.

Let us now model the tenant's decision-problem under this regime of tenancy rent control. Let us assume that all tenants have the same option (irrespective of their types) if they reject leasing an apartment.[8] They could locate in an area not covered by the rent-control law (most often suburban communities, but, as in the case of New York City and Los Angeles, sometimes rent-control coverage is only partial and rent-controlled apartments exist side-by-side with non-rent controlled apartments), buy a house, live off friends, or live in motel rooms or mobile homes, which are not covered by the rent-control ordinances. That the preferred outside option is the same across types is reasonable if one imagines that, for example, the suburban rental market is competitive. Thus rents will be set to cover costs and will be freely adjusted to inflation, neither of which depends on tenant type. Of course living in a suburb often involves additional transportation costs to those who work in the city making the option of locating to them less desirable all else equal. In a competitive environment, however, if the cost of making an apartment available to rent is the same across landlords (that is, in both rent-controlled and non-rent controlled areas), then even if renting in the non-controlled areas involves no additional costs to tenants, the results of the model will hold. The landlord's zero-profit condition assures this.

The outside option gives a person a lifetime utility of B. We assume that all tenants receive the same lifetime benefit from renting an apartment, A, and must, of course pay rent R, which, in present value terms is Rv_i for a type i tenant. We assume that $A > B$, and define the difference, $A - B$, as D. Therefore, a tenant will lease an apartment if and only if, $A - Rv_i \geq B$ or $Rv_i \leq A - B \equiv D$. What we mean by this, in operational terms, is: Irrespective of a tenant's type (which is here exogenously given), if a tenant finds that the present value of rentals exceeds D, the tenant will opt out of tenancy.

If a tenant is of type i, and the rent is R, the present value of rentals paid by the tenant is clearly Rv_i, as in (18.5), with v_i as defined by (18.1) or (18.2). Hence, a type i tenant will opt for tenancy as long as $Rv_i \leq D$. By Lemma 18.1 we know that as R increases, the shortest-staying tenants (that is, of type 1) will be the first ones to opt out of tenancy, followed by types 2, 3, and so on, with the last to opt out being the longest stayers (type n). Since the short stayers are the more attractive tenants from the landlord's point of view, this is what drives the adverse-selection process in this model.

Now consider a landlord who has one property to lease out. Let $V(R)$ be the landlord's expected present value of rental when the per-period rent is R. Following the argument in the above paragraph, we can now compute what $V(R)$ will be like as R varies. An important and interesting implication of this is the following.

LEMMA 18.3. $V(R)$ reaches its maximum when $R = D/v_n$.

Note that D/v_n is the critical rent above which the longest-staying tenants opt out of tenancy. The proof of Lemma 18.3 is obvious with the use of a somewhat unusual diagrammatic technique that we develop below. Let us first explain how $V(R)$ can be represented diagrammatically. Consider a case where $n = 3$. In Figure 18.1, the horizontal axis represents R. In this figure, draw the lines Rv_1, Rv_2, and Rv_3. By Lemma 18.1, Rv_1 is the steepest, followed by Rv_2, and then Rv_n or (what is the same here) Rv_3. Draw a horizontal line at height D from the origin and mark off the critical rents, namely, D/v_1, D/v_2, and D/v_3 where each type drops out of the rental market. All this is shown in Figure 18.1. Being a model of asymmetric information, it is not surprising that the figure is similar to the one in Stiglitz and Weiss (1981) which plots bank return as a function of the interest rate.

In the same figure, draw $Rv_{(1)}$, $Rv_{(2)}$, and $Rv_{(3)}$. Recall that $Rv_{(i)}$ is a weighted average of Rv_i, Rv_{i+1}, ...and, Rv_n. It follows that $Rv_{(3)}$ coincides with Rv_3.

Now suppose the monthly rent is below D/v_1. Then all three types of tenants seek tenancy. Hence the landlord's expected present value of rentals earned (that is, V(R)) is given by $Rv_{(1)}$. Once R exceeds D/v_1, type 1 ceases to lease in

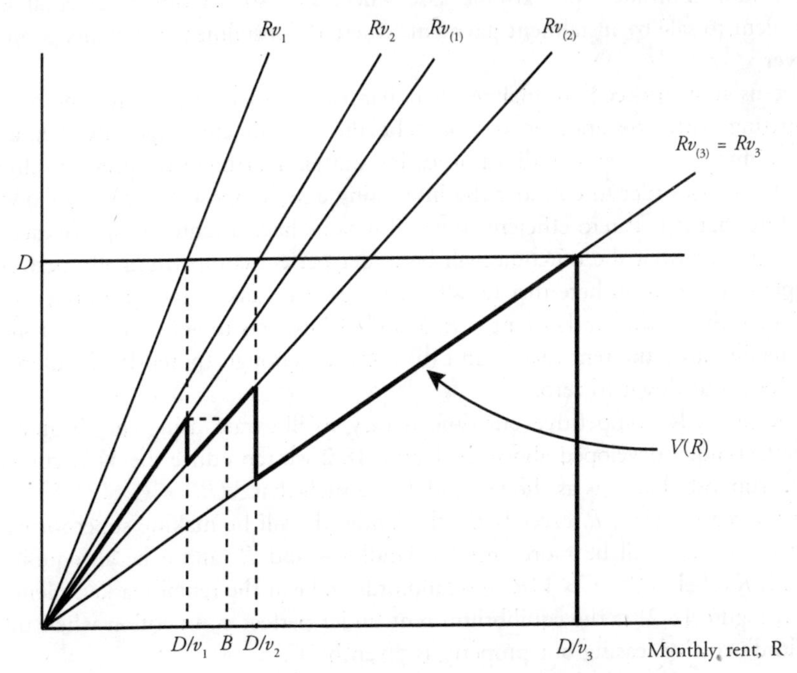

Figure 18.1 The V(R) Function

property. As long as $R \leq D/v_2$, $V(R)$ is equal to $Rv_{(2)}$. Beyond D/v_2, $V(R)$ equals $Rv_{(3)}$. Hence the landlord's expected present value of rentals, $V(R)$, must satisfy the following:

$$V(R) = \begin{cases} Rv_{(1)}, & \text{if } R \leq D/v_1 \\ Rv_{(2)}, & \text{if } D/v_1 < R \leq D/v_2 \\ Rv_{(3)}, & \text{if } D/v_2 < R \leq D/v_3 \\ 0, & \text{if } D/v_3 < R \end{cases}$$

The $V(R)$ function is illustrated by the thickened line in Figure 18.1.

One easy implication of the definitions is that $Rv_{(i)} < Rv_i$ for all $i < n$. It immediately follows that all the peaks of $V(R)$ excepting the one at rental, D/v_n will be dominated by the peak at D/v_n (as shown in Figure 18.1). This completes the proof of Lemma 18.3.

Note that if the opportunity cost of leasing property differed for each type, that is, the opportunity cost to type i was D_i instead of D, Figure 18.1 would have to be adjusted by drawing in D_1, D_2, and D_3 and locating the critical rentals D_1/v_1, D_2/v_2, and D_3/v_3. Lemma 18.3 would not then be necessarily valid. However, the inefficiency results that we prove do not hinge critically on the assumption of 'equal opportunity cost'. And so, in view of its simplifying nature, this is an assumption that we continue to use.

It is now easy to see that if the rental market is monopolistic (in the sense of there being one landlord with one property to lease out) then she would set the rent at D/V_n. All but the longest stayers would be driven out of the market in equilibrium. Consider the extreme case where $t_n = \infty$. In this case rental is equivalent to sale by instalment payment, where the instalment payments go on forever.

Let us now proceed to analyse what happens if there are many landlords competing with one another as would be the case in any large city. Let us assume that the cost to a landlord of leasing out an apartment (in present-value terms) is C. In order to consider the interesting case, suppose $C < D$. It follows therefore that it is Pareto efficient to let all tenants have a rented property each. But let us see what the outcome will be under perfect competition. By perfect competition we mean here that (a) all agents are price takers (which implies, in particular, that if the market rent is R, a landlord expects to get no tenant if she unilaterally raises the rent to R') and (b) there are enough (potential) landlords to drive profit down to zero.

The perfectly competitive outcome is easy to illustrate using the diagrammatic technique developed above. In Figure 18.2 we reproduce the $V(R)$ curve. Let us suppose that C is as shown, and R^* be such that $V(R^*) = C$.

If the market rent, R, exceeds R^*, the landlords will be making supernormal profits. So there will be more entry of landlords and R cannot be an equilibrium. If R is below R^*, $C > V(R)$ and landlords will exit the rental market. Hence R^* (in Figure 18.2) is the equilibrium rent under perfect competition when the landlord's cost of leasing out property is given by C.

Let us now consider the case where the landlord's cost, instead of being C, is

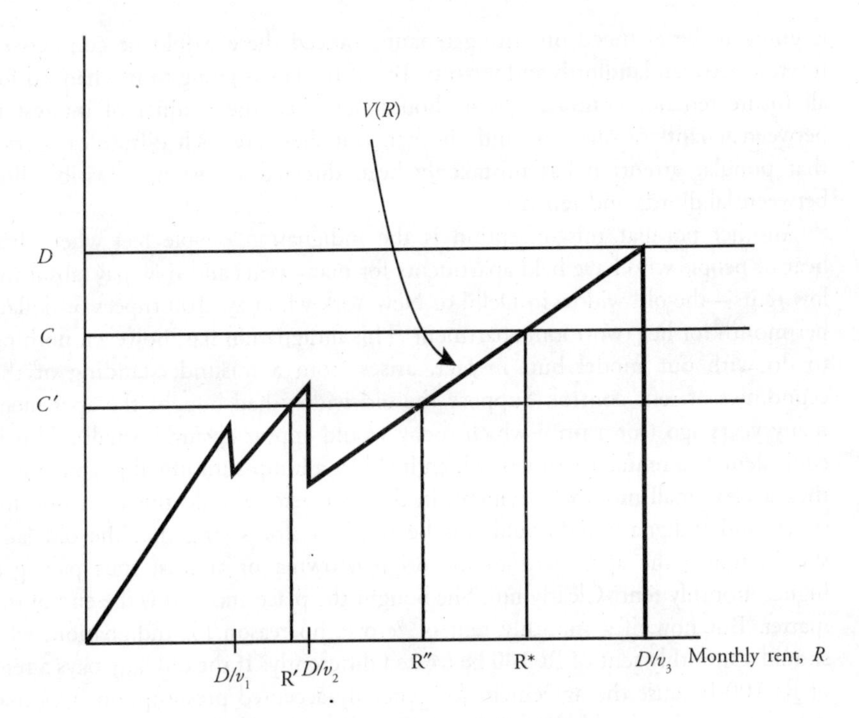

Figure 18.2 Rent Control Equilibrium

C'. Then there are two rental rates, R' and R'' at which $C' = V(R') = V(R'')$. Are both of these equilibria? It depends on how exactly we interpret a 'competitive market'. If by that we mean that landlords cannot change rents in either direction (as in Mas-Colell et al. 1995: ch. 13) then both R' and R'' constitute equilibria. However, it seems reasonable to argue that while no landlord can unilaterally deviate from the market rent in an upward direction, they can unilaterally deviate in a downward direction (without losing all tenants). Then R'' ceases to be an equilibrium. First, we explain this intuitively; and then (at the end of this section) we give it a formal game-theoretic interpretation.

The intuitive reason is simple. Suppose all landlords are charging R''. Then if one landlord cut her rent to $\hat{R} \in (R', D/v_1)$, all tenants would try to lease from this landlord and she would make a positive profit, since (as is clear from Figure 18.2) $V(\hat{R}) > C'$. Hence, R' is the only competitive equilibrium in this case.

Let us see how different people are affected in equilibrium. By comparing Figures 18.1 and 18.2, it is clear that the costs of rental to types 1, 2, and 3 are $R'v_1$, $R'v_2$, *and* $R'v_3$. By marking the R' point of Figure 18.2 in Figure 18.1 it is clear that $R'v_1 > R'v_2 > R'v_3$ and $R'v_2 > C' > R'v_3$. Hence the short stayers pay rents that are too high (above the cost to the landlords) and long stayers pay rents that are too low (type 1s do not rent at all in this market as $R'v_1 > D$). This is the real dividing line in the rent-control debate.

Most of the popular divisions arise between landlords and tenants. If the law is going to be changed on sitting tenants, indeed there would be conflicts of interest between landlords and tenants. But if the law is going to be changed for all future tenancy contracts (as it should be), then the conflict of interest is between tenants of one type and another. But these are such diffuse categories that popular attention has mistakenly been directed at the more visible line between landlords and tenants.

Another popular misconception is the indignation people feel when they hear of people who have held apartments for many years and now pay 'absurdly' low rents—the old widow in Delhi or New York who pays 100 rupees or dollars per month for her two-room apartment. This indignation has, however, nothing to do with our model but, in fact, arises from a misunderstanding of the economics of rent control. Suppose the old widow had bought the apartment many years ago (for a price which today would appear absurdly small). This is equivalent to a rental agreement which involves a lump-sum initial payment and then a very small monthly payment, in this case zero. Should this be reason for shock and indignation? Should this be the basis for saying that the old lady should return the apartment to the original owner or at least start paying a higher monthly rent? Clearly not. She bought the place and that is the end of the matter. But now, if a monthly rent of zero is no reason for indignation, why should a monthly rent of Rs 100 be treated differently? If the old lady pays a rent of Rs 100 because the agreement (or generally accepted presumption) was that that is what she would do forever, then we could, effectively, think of her as having bought the place under the agreement to pay in monthly instalments of Rs 100.[9] Of course, Rs 100 looks very small today; but zero looks even smaller and that is what a person who bought the apartment she lives in pays. Indeed in a free-contract system, someone may sign exactly the contract the old widow signed and thirty years down the road the rent will appear 'absurdly' low. This, therefore, is clearly not the essential difference between a system of rent control and one of free contract.

Return to the case illustrated in Figure 18.2 where the landlord's cost is C. In equilibrium everybody excepting the type-n tenants are driven off the market. Yet for each tenant type, i, there exists a rental such that both landlord and tenant would benefit. This implies that under rent control the competitive equilibrium can be Pareto inefficient.

If the rent-control law is revoked and replaced by a system in which a tenant and a landlord can sign on any rental and eviction contract, it is easy to see that Pareto efficiency is attained. This is because the asymmetric information problem is not insurmountable here. Though for a new tenant his type is not transparent to the landlord, at his time of departure his type gets revealed. So by writing a departure-date-contingent contract (or by putting in a rent escalation clause) landlords can get around the asymmetric information problem. The problem with most rent-control regulations is that they tend to disallow (or render illegal) clauses in rental contracts that allow landlords to overcome the asymmetric information problem.

It is also worth noting that landlords may try to counter some of the disadvantages of tenancy rent control by using harassment. If a landlord expects to benefit from a tenant vacating his unit, he will have the incentive to harass the tenant and inadequately maintain the property. This could explain why certain kinds of deposit money are discouraged under the law.

Let us, for completeness, consider the case where a landlord and a tenant can agree to any contract and the state legal machinery ensures that the contract is adhered to. Under such a legal regime, one kind of contract that would achieve optimality is a fully inflation-indexed rental contract. Suppose a landlord writes a contract where the *real* rental is R each month. In other words, the nominal rent is raised each month sufficiently to correct for the amount of inflation. Under such a rental contract, the tenant type is unimportant to the landlord because no matter who the tenant is, if the real monthly rental is R, the landlord's present value of earnings is $R/(1 - \delta)$. If the cost of leasing out is C, the competitive *real* rental rate is R^* where $R^*/(1 - \delta) = C$.

In other words, in a competitive equilibrium the nominal rent in period 1 is $(1 - \delta)\ C = R^*$. In the next month it will be $(1 - \delta)\ C / \beta$. The following month $(1 - \delta)\ C / \beta^2$, and so on. As long as $C \leq D$, all tenants will be leasing in apartments and the outcome is Pareto optimal.

If we have a rent-control law which does not allow rent-escalation clauses (at all or adequately) but does allow departure-date-contingent rentals, we once again get optimality. We could then think of the landlord offering contracts like $\tilde{R}(1)$, $\tilde{R}(2)$, ..., which say that you need make no monthly payment (it is easy to generalize and allow for *some* nominally fixed monthly payment) but if you leave the apartment after t months you make a lump-sum payment of $\tilde{R}(t)$. In that case, it is easy to see that the competitive outcome is Pareto optimal.

In reality landlords do often mimic this system. They take initial deposits from tenants and promise to return part of the money if the tenant leaves early. The returns are, however, never quite so finely-tuned as in the above paragraph, for fear of falling foul of rent-control laws.

What was described above as a competitive equilibrium may also be characterized as a Nash equilibrium of a game, as done by Mas-Colelll et al. (1995: 443–50) in the context of Akerlof's (1970) model of adverse selection. Mas-Colell et al. construct a model of Bertrand competition between two landlords, *each of whom can supply an unlimited number of apartments* at a cost of C each. The italicized part of the above sentence is clearly an unrealistic assumption. It is technically necessary because the existence of capacity constraints can give rise to well-known existence problems.

We get around this problem with a different description of the game. In our model there are m (potential) landlords where m exceeds the total number of tenants, t. Each landlord can offer at most one apartment for lease. If she does so, then she incurs a one-time cost of C. Each landlord's strategy set, S, is equal to $\{N\} \cup [0, \infty)$. If landlord i chooses N, it means she does not offer an apartment for lease. In that case she does not incur C and her profit is zero. If she chooses a strategy $R_i \in [0, \infty)$, it means she offers an apartment for lease. Her profit then depends on R_i and other landlords' choice of strategies.

We are essentially looking for a strategy m-tuple $(s_1, ..., s_m) \equiv s$ which is a refinement of a Nash equilibrium in which all the entrants choose the same strategy. We call this a 'uniform Nash equilibrium'. A rental valuc of R is a uniform Nash equilibrium if, for some $t > 0$, $m - t$ landlords choose strategy N and all the other t landlords choose the same rent $R \in [0, \infty)$ and these strategies constitute a Nash equilibrium.

The pay-off functions of the landlords are assumed to have the following properties: (a) If among all the landlords who choose to enter, j selects the smallest rent, then landlord j expects a pay-off of $V(R_j) - C$. (b) If t landlords enter and all but one of them charge the same rent R and the deviant charges $R' > R$, then the deviant's expected profit is $-C$.

(b) is a strong assumption but it mimics the idea of competition. It is a game-theoretic formalization of the concept of equilibrium intuitively used in many models of adverse selection, such as Stiglitz and Weiss (1981).

Given these assumptions, the uniform Nash equilibria of this m-person game coincide exactly with the competitive equilibria described above. In Figure 18.2 if the cost is C, the only uniform Nash equilibrium is R^* and if the cost is C', the only uniform Nash equilibrium is R'.

The only case where we can get multiple equilibria is the non-generic special case where the cost, C, is such that the horizontal line just touches a peak. That is, there exists $i < n$ such that $V(D/v_i) = C$. Barring this special case, a rental, $\tilde{R}$, is a competitive equilibrium or a uniform Nash equilibrium if

$$\tilde{R} = \min\{R / V(R) = C\}.$$

Though we have in this model assumed a perfectly elastic supply of rental accommodation, this is not necessary. The method of analysis developed here can be extended to the case where the supply of rental accommodation is perfectly inelastic or upward rising with respect to the return on rental. For reasons of illustration consider the case where the supply of rental accommodation is fixed at a number S, and C happens to be zero. Let us consider the case illustrated in Figure 18.1. Again, let the total number of (potential) tenants be t. Hence, there are tp_i number of tenants of type i. To illustrate the analysis with a special case, suppose $S = tp_2 + tp_3$. We know immediately that the only rents that can possibly qualify as equilibria are the ones in the interval $[D/v_1, D/v_2]$ in Figure 18.1 since only in such cases demand equals supply. If however the rent is too close to D/v_1, in particular less than B, each landlord will have an interest in under-cutting others for the kind of reason discussed above. Hence, with a fixed supply of rental accommodation, all rents in the interval $[B, D/v_2]$ comprise equilibria. By considering the case where $S \neq \{tp_3, t(p_2 + p_3), t\}$, we can create equilibria where markets do not clear but this happens in the special sense of tenants, who are indifferent between leasing an apartment in the rent-controlled area and the outside option, being excluded from the rent-controlled sector.

18.3. Inflation and Rents

Our model can be used to derive an interesting testable proposition concerning the relation between inflation rates and rents in rent-controlled areas. The analysis, however, turns out to be more complex than it seems at first sight.

It is intuitively obvious that as the rate of inflation rises, (1) the inefficiencies of a rent-controlled regime become more acute and (2) the long stayers become *more* unattractive as tenants from the point of view of the landlords than the short stayers. (1) and (2) are technical results; we derive these formally in Lemma 18.4 and Lemma 18.5, respectively. It is an implication of these two lemmas that a higher inflation rate will result in a higher *real* rental rate for starting tenants. Roughly speaking, the logic behind this proposition is that as inflation picks up, the distortions created by tenancy rent control get exacerbated; the return earned by landlords drops; and the market responds by causing rents to rise. This may well result in more tenants being excluded from the rent-controlled market. In a country like India where the entire urban rental market is controlled, this means more tenants will be excluded from the urban rental market altogether.

Let us postpone further discussion until after the formal results are derived.

Note that in our model, a rise in the inflation rate simply means a fall in β, with zero inflation being equivalent to $\beta = 1$. Throughout the following analysis we will assume $\beta \leq 1$, that is, we exclude the case of negative inflation. To make explicit the dependence of returns on the rate of inflation, we will here write v_i and $v_{(i)}$ as $v_i(\beta)$ and $v_{(i)}(\beta)$, respectively.

Lemma 18.4

$$\text{If } \beta' < \beta, \text{ then } v_{(i)}(\beta') < v_{(i)}(\beta)$$

Lemma 18.5

$$\text{If } \beta' < \beta, \text{ then } \frac{v_{(i)}(\beta')}{v_i(\beta')} < \frac{v_{(i)}(\beta)}{v_i(\beta)}.$$

Using the two lemmas, it is now possible to derive a testable proposition concerning the relation between inflation and rent. There is no reason to believe that inflation alters the outside options of agents in any systematic way. So we will take it to be neutral. Hence D and C remain unchanged as the inflation rate changes. Let us also work here with the assumption of 'free entry' of landlords, though that can be altered along the lines of analysis in the last part of Section 18.2.

Let us see what happens to the $V(R)$ curve as the rate of inflation rises, that is β changes to β', which is less than β. From Lemma 18.1 we know that the ray $Rv_{(i)}$ will now be flatter for every i.

Next consider each peak of the $V(R)$ line, for instance, the peak at D/v_2 in Figure 18.1. Clearly, the height of that peak is given by $Dv_{(2)}/v_2$. From Lemma 18.5 we know that as inflation increases (that is, β drops) this peak must get lower. This is, of course, true for every peak.

From the analysis in the previous section we know that the equilibrium point on the $V(R)$ curve must be at a point where the horizontal line at height C first hits the $V(R)$ curve from the left. From observations in the above two paragraphs we know that if the inflation rate increases, the $V(R)$ curve will shift right with the new peaks no higher than before. It follows that the equilibrium monthly rent, R, will rise. Since the equilibrium is the real rent faced by a starting tenant, what we have just established is the proposition that, as the inflation rate rises, the *real* rent for new leases goes up in economies in which there is tenancy rent control.

It is also easy to see that if the rise in inflation is sufficient, the rise in the real rental can be so sharp as to exclude a whole class of tenants who were earlier leasing apartments from the rental market. This is not surprising at all because the inefficiency of tenancy rent control gets more acute as the inflation rate rises.

18.4. Endogenous Quit Decisions

There are several directions in which one can modify and extend the above model. We will, in this section, consider one involving the endogenization of the tenant's type. It is true that some people are inherently prone to moving and some have transferable jobs. But no matter what the inherent penchants, people do modify their behaviour depending on the conditions in the rental market. If inflation is very high and a rent-control order holds the nominal rent constant for sitting tenants, inherently peripatetic individuals may try to change their ways and stay put in one place, and some people with transferable jobs may quit such jobs. In this section we shall try to show that the endogenization of tenant types can generate some very interesting results, including the generic possibility of multiple equilibria. Moreover, the removal of rent control can result in a uniform lowering of rents.

Let us consider the case where all tenants are innately identical but they can *choose* to be one of two types: 1 and 2. The assumption *of* ex ante identity is inessential and is made for ease of explanation. Type i changes his apartment every t_i months where $t_1 < t_2$. In other words, a tenant *chooses* to be a short stayer or a long stayer. Again, for reasons of simplicity, let us assume $t_2 = \infty$. In other words, a potential tenant has to decide whether to be a short stayer or settle down permanently in a rented apartment. Let us see what happens if we have the kind of tenancy rent control discussed in the previous section.

Consider alternative life strategies for the tenant. If a tenant decides on a career path in which he moves to take up a better job, wherever such opportunities arise, he will be a short stayer. Let his expected lifetime wage income in this case be W_1. If alternatively he chooses a life where he stays in one place and takes up whatever job he gets in the vicinity, he is a long stayer and his expected lifetime wage-income is W_2. We assume $W_1 > W_2$.

Suppose the market rent is R. A person who decides to be a tenant will choose to be a short stayer if and only if $(W_1 - W_2) \geq (v_1 - v_2)$. By Lemma 18.1, we know that the right-hand term is positive. Hence, there exists a critical rent size, $\bar{R}$, such that if $R \leq \bar{R}$, tenants prefer being of type 1 and if $R > \bar{R}$, tenants prefer to be of type 2. Clearly,

$$\bar{R} \equiv \frac{W_1 - W_2}{v_1 - v_2} \tag{18.7}$$

Let us, as in the end of Section 18.2, describe the outcome of the rental market by thinking of this as a game among the m landlords. As before $m > t$, where t is the number of potential tenants. For simplicity assume D is very high; so the potential tenants always choose to be tenants. A tenant's crucial decision now pertains to what type he will be. If all landlords charge the same rent, R, each tenant's choice has already been described. If $R \leq \bar{R}$, each tenant chooses to be of type 1. Otherwise he is of type 2.

Let us denote this decision by the function T: $[0, \infty) \rightarrow \{1, 2\}$. $T(R) = 1$ if and only if $R \leq \bar{R}$. Thus $T(R)$ tells us what type the tenants will be, if there is only one rent, R, prevailing in the market. Now, consider the case where, with all the landlords charging R, one landlord deviates to R'. What can the deviant landlord expect? As in Section 18.2, we assume that if $R' > R$, she expects to find no tenants. If $R' < R$, tenants will of course come to her, but the question is what lifestyle will the tenant choose: short staying or long staying? It seems to us, and this is what we will assume, that all tenants will be of type $T(R)$—even the tenant who rents at rate R' from the deviant landlord. More formally, in the language of games used at the end of Section 18.2, we assume: (c) If t landlords enter the rental market and all but one of them charges a rent of R and one landlord charges a rent $R' < R$, then the tenants attracted by the deviant landlord will be of type $T(R)$.

This assumption seems to be intuitively very reasonable. Suppose you live in a city with a tenancy rent-control law in which every landlord, except one, charges a very high rent. The one exception is your landlord who charges a low rent of R'. Suppose if *every* landlord charged R' you would have adopted the short-staying lifestyle (involving moving every time you got a better job). What will you do when only your landlord charges R'? There seems little rational motivation for you to adopt the short-staying lifestyle. In fact, since you know that higher rents exist everywhere else, you will have an extra reason to stay put where you are.

This assumption is crucial to our model and it is worth clarifying that it is based on intuition, which is external to our model. We, however, believe that it is a tenable assumption, and hope that future work will be able to derive this postulate (c), from more basic axioms.

Figure 18.3 considers the case in which all landlords charge the same rent R and the thick line shows each landlord's expected lifetime rental income, $V(R)$. Note that if $R \leq \bar{R} \equiv (W_1 - W_2) / (v_1 - v_2)$, all tenants are of type 1 and if $R > \bar{R}$, all tenants are of type 2.

Now suppose (as in Section 18.2) a landlord's cost of leasing out an apartment is C (as shown in the figure). To fix our attention on the interesting case, we consider one where the horizontal line at C intersects $V(R)$ more than once (at rents R_L and R_H). Unlike in Section 18.2, here both R_L and R_H constitute competitive equilibria or, equivalently, uniform Nash equilibria. It is obvious that R_L is an equilibrium. So consider the case where the market rent is R_H. Landlords' profits are zero, but it seems, at first sight, that an individual landlord can do better by charging a lower rent—anything between R_L and $\bar{R}$. Suppose one landlord does so and charges $R \in (R_L, \bar{R}]$. She will have no problem getting a tenant of course. However, the tenant who moves in will not behave like a type 1 tenant, because if he gives up this tenancy there is no reason for him to expect that he will again find an apartment for rent R. Hence, the deviating landlord's expected profit will be $Rv_2 - C$. This is non-positive for $R \in (R_L, \bar{R}]$ .So no one benefits from deviating from R_H, which is a competitive equilibrium. For a formal game-theoretic argument we have to merely cite assumption (c) above to explain why it does not pay to undercut.

The argument that explains the possibility of multiple equilibria given tenancy rent control is based on the assumption that there are limits to the number of apartments a single landlord can offer (for simplicity assumed to be one in this chapter). If a landlord could undercut R_H by offering $R' \in (R_L, \bar{R})$ and supply a large number of apartments at that rent, she may be able to cause

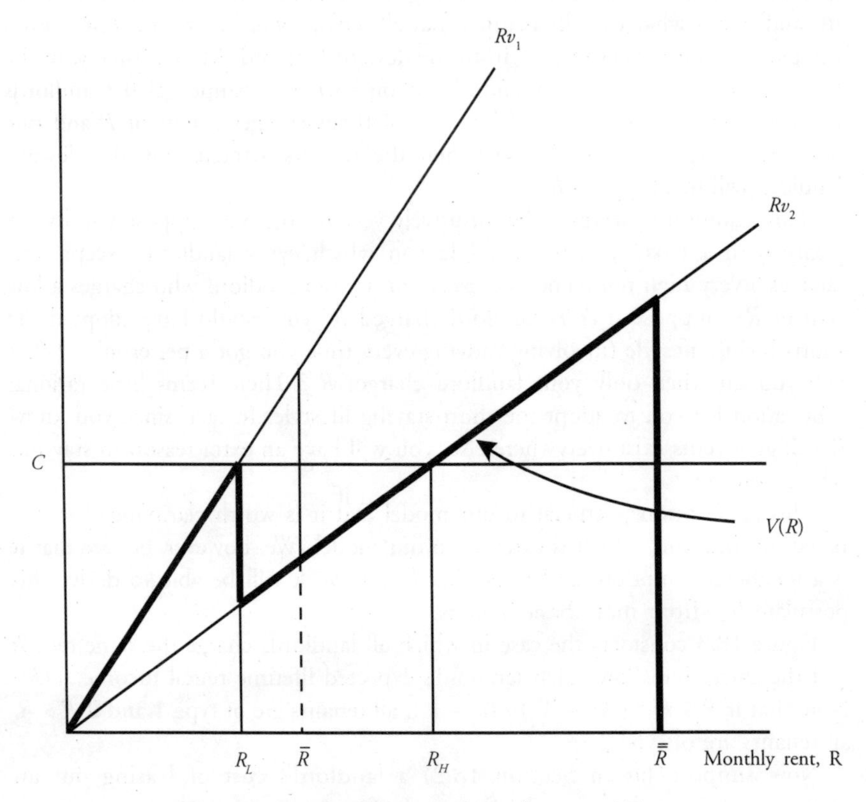

Figure 18.3 Equilibria with Endogenous Types

tenants to alter their life's strategy from being long stayers to short stayers. In other words (c) would then no longer hold, and so R_H would cease to be a uniform Nash equilibrium. But in any metropolitan city it does seem reasonable to assume that no single landlord can cause such a large infusion of apartments as to induce tenants to alter their life strategy.

Finally, if there is no rent control, clearly then R_H will cease to be an equilibrium, because landlords can write type-contingent contracts. So a landlord can deviate to a rent such as $\bar{R}$ and make it contractual on the tenant quitting after t_1 periods. It is easy to see that the only equilibrium is at R_L in Figure 18.3.

The upshot is that, under tenancy rent control, both R_H and R_L are equilibria, whereas under the free-contract system only R_L is. Suppose rent control is in effect and it is the bad equilibrium, that is, R_H, that prevails. In comparison to this, no rent control is not only Pareto efficient but it is Pareto dominant. The removal of rent control will result in all rents going down.

18.5. Extensions: Turnover Costs, Signals, and Identity

In this section we explore two extensions to our model: one in which the landlord incurs turnover costs, and the other where landlords are able to tell something about the expected tenure of a potential tenant by identifying the group to which the potential tenant belongs.

Turnover costs are common in the rental housing market and typically consist of cleaning, repainting, and renewing worn out appliances upon vacancy. Let us consider the case where there exists a fixed turnover cost, which the landlord has to pay each time a tenant vacates an apartment. In this case, every time a tenant leaves, the landlord has to incur a transaction cost of ϕ (the qualitative results are unchanged whether the turnover costs occur at the start of a tenancy or the end). Then in (18.1) we need to add the term $-\ \delta^{t_i}\phi$ to the right-hand side and that will mean (18.2) would be:

$$\hat{v}_i = \frac{1-(\beta\delta)^{t_i}}{(1-(\delta^{t}))\,(1-\beta\delta)} - \frac{\delta^{t_i}\phi}{1-\delta^{t_i}} \tag{18.8}$$

The hat on the v_i is to mark out this general case. In other words, $\hat{v}_i$ is the present value of income earned by a landlord if she charges a nominal monthly rent of $1 in real terms for every new tenant, gets only tenants of type i, and incurs a cost of ϕ every time a tenant quits. We define the cost term of (18.8) as:

$$\Phi_i \equiv \frac{\delta^{t_i}\phi}{1-\delta^{t_i}} \tag{18.9}$$

It is easy to see that $\Phi_i < \Phi_j$, for all $i > j$. Therefore, there is now a tension between the benefits to the landlord from a short stayer, that rents are not eroded as much as with a long stayer, and the costs, that turnover costs are incurred more frequently. This can lead to even more types of tenants being kept out of the market. Define $\Phi_{(i)}$ as expected present turnover costs to the landlord when all tenants of type i or above make themselves available to the landlord as tenants. We can then write $\hat{v}_{(i)}$ as $v_{(i)} - \Phi_{(i)}$. If the rent is R and only tenants of type i and higher seek tenancy, the landlord's present value of income from leasing out one apartment is given by $Rv_{(i)} - \Phi_{(i)}$. This follows from (18.6), (18.9), and (A18.3). It is now easy to see that with this specification it is no longer generally true that if $i < j$, $\hat{v}_{(i)} > \hat{v}_{(j)}$, and so from the landlord's point of view the shortest stayers are no longer necessarily the ideal tenants.

The tension between the desirability of oft-reset rents and undesirability of frequent turnover costs can potentially alter the equilibrium from Section 18.2. Shorter stayers are likely to be excluded from the rental market by landlords who want to avoid frequent turnover costs. They accomplish this by setting rents high enough so that short-staying tenants exit the market. Therefore, turnover costs can lead to the exclusion of tenant types that would not have been excluded in the absence of these costs.

In reality tenants are not as faceless as assumed in the model. People differ by race, age, profession, nationality, and so on. Some of these may give a potential landlord a hint of how long or short stayer the tenant is likely to be. A person of a different nationality with a temporary visa cannot be a long stayer. Landlords may trust people of a certain age group or a certain race more. In that case, at the time of leasing they can ask such tenants how long they plan to stay and put greater trust in the answer than they would with other kinds of tenants. Landlords may also believe that certain groups of tenants are more likely to leave when asked to do so. They will then prefer tenants belonging to such a group.

The examples that we have given above are a mixed basket of inherent characteristics and signals which a tenant can choose (example, his profession). For reasons of simplicity let us consider the case of inherent characteristics. Suppose people belong to several different races, *A, B, C*, ... The expected length of stay of a tenant differs depending on which type one is and this is known to landlords. So though, within each category, there are short stayers and long stayers, on average *A* may be the shortest stayer; followed by *B, C*, and so on. In such a market what we will expect to see is a variety of rents. These will be contingent on the tenant's race.

Thus, our model predicts that people in a profession which is known to involve transfer and moving, or belonging to some identifiable group that is known to be footloose, will be found paying a lower rent than those who are known to sink roots in one place and settle down.

18.6. Conclusion

Rent-control laws have been enacted in many countries around the world, making them one of the most popular public policy prescriptions among metropolitan governments. Unfortunately, knowledge of the effects of tenancy rent control (which is one of the most pervasive forms of rent control) is inadequate, especially in the context of positive inflation. This chapter constructed a model of tenancy rent control and showed that this kind of rent-control system, with asymmetric information and exogenously given tenant types (the 'type' of a tenant being identified in terms of how long a tenant expects to stay in the same apartment), lead to outcomes that are Pareto sub-optimal. Free contracting, however, allows the agents in this model to overcome the asymmetric information problem. The chapter then studied a model in which how long a tenant stays in one place is decided by the tenant on the basis of market signals. This captures the fact that many agents make lifestyle choices depending on the conditions of the rental housing market. Endogenizing the tenant's type gives rise to the possibility of multiple equilibria in our model. Removal of rent-control laws can not only increase efficiency in the rental market, but can also lead to a general lowering of rents, making all tenants better off.

A number of empirical implications arise from our model. Since landlords cannot write departure date-contingent contracts or have a rent-escalation clause included in the contract, the landlord must set initial rents higher to compensate for the erosion of real rents suffered during occupancy. This should lead to across-the-board higher rents in rent-controlled apartments that are being offered on the market (vacant apartments) than comparable offerings in non rent-controlled cities (as found in Nagy 1997). One would also expect to find evidence of a tenure discount in rent-controlled cities (as in Nagy 1997; and Börsch-Supan 1986), where tenants who have rented the same apartment for many years pay considerably less in rent than do tenants who have only just recently taken possession of an apartment.[10] Some recent research (see Olsen 1990) suggests that the construction of new housing need not be any less frequent in a rent-controlled city. However, in our basic model, as certain types are excluded from the rental housing market, the supply of rental housing in the rent-controlled market is likely to fall. In addition, if increased inflation were severe enough to cause the exclusion of even more types (as in Section 18.3), then the supply of rental housing would fall as well.

An important implication of the model is that rent control might decrease the mobility of the labour force. As sitting tenants are reluctant to move from a rent-controlled apartment, they are less likely to accept a higher-paying job in another city. Therefore, empirically, we would expect to find that the average tenure of renters is higher in rent-controlled cities (as in Nagy 1997; and, to some extent, in Olsen 1990), and that rent control reduces tenant mobility (as in Nagy 1995; and Ault et al. 1994). In fact, Nagy (1997) presents empirical findings that are directly in line with our model. He studies New York City apartments under the post-1974 rent-stabilization scheme and finds that in 1981, rent-stabilized apartments had higher initial rents than non-stabilized apartments. Six years later, for those tenants who remained in the same apartment in both sectors, those in the stabilized sector paid lower rents. In addition, tenants in the stabilized sector had longer tenure durations than tenants in the non-stabilized sector. Thus, the received empirical evidence generally supports these hypotheses, drawing a picture remarkably similar to the one that is implied by our model, but the scarcity of detailed empirical evidence suggests that there is still work to be done in this area.

Our model also drew attention to some systematic relations between rates of inflation and rental rates for property under tenancy rent control. These are testable propositions but evidence on this is hard to come by. It is hoped that our model, by clarifying the theoretical link between inflation and rent, will prompt researchers to collect and verify empirically the claims that arise from this theory.

From the above set of results, it is easy to get the idea that the optimal policy solution is to free the rental housing market of all government restrictions. We caution the reader from extending this logic too far, however. As we discussed earlier in the chapter, free contracting in the rental housing market, in the sense that we use it, does entail certain important responsibilities on the part of government. The government provides the framework in which contracts are enforced, and though in our model the absence of rent control was associated with a system of free contract, there will in reality be three important kinds of limits on the range of contracts allowed. First, since every society considers certain kinds of activities illegal, a contract that specifies the use of some illegal activity would naturally not be recognized even if both parties voluntarily agree to it. A contract which entails the landlord killing a tenant who fails to pay the rent would belong to this category. Second, a contract which adversely affects an uninvolved 'third party' (that is, someone who is not a signatory to the contract) would be considered illegal. Finally, what one has to keep in mind is that, in this age of lawyers, contracts can soon become so complicated that it becomes virtually impossible for the signatories to understand fully what exactly they are agreeing to. In such a scenario, the more savvy can 'trap' the others into transactions that they would not have got into *if* they understood the full ramifications of the contract. To prevent this from happening, a practical response is to set some broad limits to the range of permitted contracts. To the extent that rent controls are themselves partly the consequence of a well-meaning attempt to restrict the range of contracts, one has to exercise common sense when limiting the terrain of possible contracts. What our model suggests is that the class of possible contracts should include rent-escalation clauses, tenancy termination clauses, and in general, contracts involving terms which are departure date contingent. This freedom of choice allows tenants and landlords to overcome the asymmetric information problem and reaches not only Pareto efficiency but may result in across-the-board lower housing rentals than what occur in the presence of rent control.

Appendix 18.1 Proofs of Lemmas

Proof of Lemma 18.1. We shall, without loss of generality, assume that $t_j = t_{i+1}$. Let v_j^k be the present value of rents earned by a landlord whose first k tenants are of type i and all others of type j. (Hence, $v_j^0 = v_j$)

We first show that $v^1_i > v_j$. Clearly

$$v^1_j = 1 + \beta\delta + (\beta\delta)^2 + \ldots + (\beta\delta)^{t_i - 1} + \delta^{t_i} v_j. \tag{A18.1}$$

Since $t_j = t_{i+1}$, and given (18.3) and (18.1), we have

$$v^1_j - v_j = \delta^{t_i} v_j - (\beta\delta)^{t_i} - \delta^{t_i + 1} v_j$$

$$= \delta^{t_i}[(1 - \delta)v_j - \beta^{t_i}]$$

$$= (1 - \delta)\delta^{t_i}\left(v_j - \frac{\beta^{t_i}}{1 - \delta}\right)$$

It is easy to see

$$v_j > \frac{\beta^{t_i}}{1 - \delta} \tag{A18.2}$$

The right-hand term is the present value of the stream $[\beta^{t_i}, \beta^{t_i}, \ldots]$, while v_j is the present value of the sequence $[1, \beta, \beta^2, \ldots, \beta^{t_i}, 1, \beta, \beta^2, \ldots, \beta^{t_i}, 1, \ldots]$. The latter sequence dominates the former, term by term. Hence, (A18.2) is true, and, therefore, $v^1_j > v_j$.

It is easy to check, $v^k_j > v^{k-1}_j$, $\forall k$, and that $\lim_{k\to\infty} v^k_j = v_i$. It follows that $v_i > v_j$.

PROOF OF LEMMA 18.2. Note that, for all k,

$$v_k = 1 + \beta\delta + (\beta\delta)^2 + \ldots + (\beta\delta)^{t_k - 1} + \delta^{t_k} v_k$$

or

$$1 + \beta\delta + (\beta\delta)^2 + \ldots + (\beta\delta)^{t_k - 1} = (1 - \delta^{t_k}) = v_k.$$

Substituting this in (18.4) , we get

$$v_{(i)} = \frac{\sum_{k=i}^{n} \left(\frac{p_k}{\sum_{j=i}^{n} p_j}\right)(1 - \delta^{t_k})v_k}{1 - \sum_{k=i}^{n}\left(\frac{p_k}{\sum_{j=i}^{n} p_j}\right)\delta^{t_k}}$$

or

$$v_{(i)} = \frac{\sum_{k=i}^{n} p_k (1 - \delta^{t_k})v_k}{\sum_{j=i}^{n} p_j - \sum_{k=i}^{n} p_k \delta^{t_k}} \tag{A18.3}$$

It is worth noting that if the term v_k were not there on the right-hand side of (A18.3) then the right-hand side would be equal to 1. Hence, $v_{(i)}$ is clearly a weighted average of v_i, v_{i+1}, ..., and v_n. It is also evident that if $j > i$, $v_{(i)}$ is obtained from $v_{(j)}$ by redistributing some of the weights away from v_j, ..., v_n to v_i, ..., v_{j-1}. Since, for all $k < j$, $v_k > v_j$ (by Lemma 18.1), it follows that $v_{(i)} > v_{(j)}$.

PROOF OF LEMMA 18.4. From (18.1) it is obvious that

$$(1 - \delta^{t_k}) = v_k(\beta) = 1 + \beta\delta + (\beta\delta)^2 + \ldots + (\beta\delta)^{t_k - 1} \tag{A18.4}$$

Writing the expression for $v_{(i)}$ derived in the proof of Lemma 18.2, with the dependence on β made explicit, we have:

$$v_{(i)}(\beta) = \frac{\sum_{k=i}^{n} \left(\frac{p_k}{\sum_{j=i}^{n} p_j} \right) (1 - \delta^{t_k}) v_k(\beta)}{1 - \sum_{k=i}^{n} \left(\frac{p_k}{\sum_{j=i}^{n} p_j} \right) \delta^{t_k}} \tag{A18.5}$$

By inspecting (A18.4) it is obvious that $\beta' < \beta$ implies $v_k(\beta') < v_k(\beta)$, for all k. From (A18.5) it follows that $v_{(i)}(\beta') < v_{(i)}(\beta)$.

PROOF OF LEMMA 18.5. Assume $\beta' < \beta$. From (A18.5) it is clear that if

$$\frac{v_k(\beta')}{v_i(\beta')} < \frac{v_k(\beta)}{v_i(\beta)}, \quad \text{for all } k > i \tag{A18.6}$$

then the proof of the lemma is complete. The remainder of this proof is therefore devoted to establishing (A18.6).

Let $k > i$. Define $m \equiv t_k - t_i$. Clearly $m \geq 1$. From (18.2) it follows that

$$\frac{v_k(\beta)}{v_i(\beta)} = \frac{1 - \delta^{t_i}}{1 - \delta^{t_i + m}} \cdot \frac{1 - (\beta\delta)^{t_i + m}}{1 - (\beta\delta)^{t_i}} \equiv Z \cdot \frac{1 - (\beta\delta)^{t_i + m}}{1 - (\beta\delta)^{t_i}}.$$

By differentiating the right-hand term with respect to β it can be checked that as β falls, the term becomes smaller. Hence, (A18.6) is proven.

NOTES

1. Arnott (1995) discusses the history of rent control in the United States and Europe and provides a useful bibliography.
2. In Los Angeles annual rent increases for sitting tenants are limited to 7 per cent, but once vacated, a new rent can be freely chosen.
3. See Linneman (1985) for a good discussion of the history of New York City rent control.
4. 'Deal is Achieved as Rent Laws Expire' by James Dao, *The New York Times,* 16 June 1997.
5. See Jarett and McKee (1997) for anecdotal evidence of the rent-increasing tactics of New York City landlords as well as a brief history of rent control in New York City.
6. Cinque (1997) discusses the non-rent protections afforded tenants by New York City's various rent-control laws.
7. In our framework we are taking this environment to be static and thus we present a static model, however it is important to note that in the dynamics of the model short stayers will appear more frequently on the market for rental housing than will long stayers. What this means, in effect, is that the present value of the returns to the landlord when facing a mix of indistinguishable potential tenants should not depend on the proportion of types in the economy [as in (18.3)] but rather the proportions of each type that are in the market for a vacant apartment at the steady state. While our method is a simplification, using the steady-state proportions will yield the same asymmetric information results as we derive using our method,

however a dynamic model of this situation would be a worthwhile undertaking in the future.

8. As explained later, much of our results would continue to hold without this assumption; but it is a useful simplifying assumption and will be used throughout.
9. There may be inefficiencies arising in this case because the original contract was unclear or because a tenant has limited rights compared to an owner (Basu 1989) but those are not our concern in this chapter.
10. Tenure discounts may also occur in non-rent controlled areas as well, however the empirical evidence is mixed. See Guasch and Marshall (1987).

REFERENCES

Akerlof, G. 1970. 'The Market for Lemons': Quality Uncertainty and the Market Mechanism. *Quarterly Journal of Economics* 84: 488–500.

Arnott, R. 1995. 'Time for Revisionism on Rent Control?'. *Journal of Economic Perspectives* 9: 99–120.

Ault, R.W., J.D. Jackson, and R.P. Saba. 1994. 'The Effect of Long-term Rent Control on Tenant Mobility'. *Journal of Urban Economics* 35: 140–58.

Basu, K. 1989. 'Technological Stagnation, Tenurial Laws, and Adverse Selection'. *American Economic Review* 79: 251–5.

Börsch-Supan, A. 1986. 'On the West German Tenants' Protection Legislation'. *Journal of Institutional and Theoretical Economics* 142: 380–404.

Cinque, R. A. 1997. 'Non-rent Protections under Rent Control and Rent Stabilization'. *New York Law Journal* 5 June.

Drier, P. 1997. 'Rent Deregulation in California and Massachusetts: Politics, Policy, and Impacts'. Paper presented at the 'Housing '97' Conference at New York University, 14 May.

Epple, D. 1998. 'Rent Control with Reputation: Theory and Evidence'. *Regional Science and Urban Economics* 28: 679–710.

Fallis, G. 1988. 'Rent Control: The Citizen, the Market and the State'. *Journal of Real Estate Finance and Economics* 1: 309–20.

Frankena, M. 1975. 'Alternative Models of Rent Control'. *Urban Studies* 12: 303–8.

Guasch, J.L. and R.C. Marshall. 1987. 'A Theoretical and Empirical Analysis of the Length of Residency Discount in the Rental Housing Market'. *Journal of Urban Economics* 22: 291–311.

Hubert, F. 1995. 'Contracting with Costly Tenants'. *Regional Science and Urban Economics* 25: 631–54.

Hubert, F. 1996. 'Rental Contracts, Endogenous Turnover and Rent Volatility'. Diskussionbeiträge des Fachbereichs Wirtschaftswissenschaft der Freien Universität Berlin.

Jarett, S. and M. McKee. 1997. *Rent Regulation in New York City: A Briefing Book.* New York: Community Training and Resource Center.

Linneman, P. 1985. 'The Effect of Rent Control on the Distribution of Income among New York City Renters'. *Journal of Urban Economics* 22: 14–34.

Mas-Colell, A., M.D. Whinston, and J.R. Green. 1995. *Microeconomic Theory.* Oxford: Oxford University Press.

Nagy, J. 1995. 'Increased Duration and Sample Attrition in New York City's Rent Controlled Sector'. *Journal of Urban Economics* 38: 127–37.

Nagy, J. 1997. 'Do Vacancy Decontrol Provisions Undo Rent Control?'. *Journal of Urban Economics* 42: 64–78.

Olsen, E.O. 1990. 'What is Known about the Effects of Rent Controls?'. Consulting Report for US Department of Housing and Urban Development, Contract Number NA 90–7272, September.

Raymon, N. 1983. 'Price Ceilings in Competitive Markets with Variable Quality'. *Journal of Public Economics* 22: 257–64.

Smith, L.B., K.T. Rosen, and G. Fallis. 1988. 'Recent Developments in Economic Models of Housing Markets'. *Journal of Economic Literature* 26: 29–64.

Stiglitz, J.E. and A. Weiss. 1981. 'Credit Rationing in Markets with Imperfect Information'. *American Economic Review* 71: 393–410.

Sweeney, J.L. 1974. 'Quality, Commodity Hierarchies, and Housing Markets'. *Econometrica* 42: 147–67.

Index